Fernando A. Bataghin
José Salatiel R. Pires
Fábio de Barros

Vascular epiphytism in the Sorocaba/Médio Tietê River Basin

Fernando A. Bataghin
José Salatiel R. Pires
Fábio de Barros

Vascular epiphytism in the Sorocaba/Médio Tietê River Basin

Conservation in Forest Fragments

Imprint
Any brand names and product names mentioned in this book are subject to trademark, brand or patent protection and are trademarks or registered trademarks of their respective holders. The use of brand names, product names, common names, trade names, product descriptions etc. even without a particular marking in this work is in no way to be construed to mean that such names may be regarded as unrestricted in respect of trademark and brand protection legislation and could thus be used by anyone.

Cover image: www.ingimage.com

This book is a translation from the original published under ISBN 978-613-9-62665-6.

Publisher:
Sciencia Scripts
is a trademark of
Dodo Books Indian Ocean Ltd. and OmniScriptum S.R.L publishing group

120 High Road, East Finchley, London, N2 9ED, United Kingdom
Str. Armeneasca 28/1, office 1, Chisinau MD-2012, Republic of Moldova, Europe
Printed at: see last page
ISBN: 978-620-7-74929-4

SUMMARY

SUMMARY 3
CHAPTER 1 5
CHAPTER 2 21
CHAPTER 3 23
CHAPTER 4 30
CHAPTER 5 196
BIBLIOGRAPHICAL REFERENCES 205

"I have the impression of having been a child playing by the sea, having fun discovering a smoother pebble or a more beautiful shell than the others, while the immense ocean of truth remains mysterious before my eyes."

Isaac Newton

To

Santo, Donatila and Marcela

SUMMARY

Around 10% of all vascular plants in the world are epiphytic. Vascular epiphytes establish themselves on host trees (phorophytes), without emitting haustorial structures, in order to gain greater access to sunlight, even at the expense of humidity conditions. This study characterized the floristics, structure and spatial distribution of the epiphytic vascular community in protected and unprotected forest fragments in the Sorocaba/Médio Tietê river basin (22°30' to 23°45' S, and 48°15' to 47°00' W). Given its size, the watershed studied has peculiar phytosociological conditions. Three types of vegetation were recorded along it, which allowed it to be divided into a downstream area, a central area and an upstream area, according to the vegetation characteristic of each area. The floristic survey was carried out in 21 sample sites (forest fragments), seven in each of the areas. The epiphytic vascular component was assessed quantitatively in 1080 phorophytes with a DBH $\geq$ 20 cm and distributed over 12 sites (90 phorophytes each site and four sites in each basin area). The forophytes were divided into six vertical strata in which the abundance of epiphytic species was recorded. A total of 176 epiphytic species belonging to 66 genera and 14 families were recorded in the watershed. Shannon's diversity index for the basin was H' = 3.695, equability (J) was 0.713 and Margalef's richness (d) was 18.39. In the Downstream Area, 56 species, 28 genera and nine families were found, with Shannon index H' = 2.948, equability J = 0.732 and Margalef richness d =6.470. In this same area 64% of the species were anemochoric and 36% zoochoric and the abundance of epiphytes did not vary significantly ($p > 0.05$) between the Core Site (UC) and its replicas, however there was a significant difference ($p < 0.05$) between the richness observed in the Core Site and its Replicas, except for the Replicica II Site. In the Downstream Area, the shape of the vertical distribution of vascular epiphytes varied significantly ($p < 0.05$) only between the Core Site and Replica Site I. For the Central Area, 64 species, 32 genera and nine families were recorded. The Shannon index was H' = 2.872, the equability J = 0.686 and the Margalef richness index (d) was 7.605. Observation of the dispersal syndrome indicated that 56% of the species were anemochoric and 44% were zoochoric. There was a significant difference ($p < 0.05$) between the richness of the Core Site and the Replica II and Replica III Sites, but there was no significant variation ($p > 0.05$) between the abundances of the Core Site and its replicas or between the shape of the vertical distribution of the vascular epiphytes of the Core Site and its three Replicas in the Central Area. In the Upland Area, 139 species, 61 genera and 14 families were found, with a Shannon diversity index of H' = 3.659, equability J = 0.742 and Margalef richness d= 16.17. In this Upland Area, 68% of the species showed anemochoric dispersal and 32% showed zoochoric dispersal. The abundance of vascular epiphytes did not vary significantly (p > 0.05) between the Core Site and its replicas, while the richness varied significantly (p < 0.01) between the Core Site and all its replicas. There was a significant difference between the shape of the vertical distribution of the vascular epiphytes that occur in the Core Site of the Upland Area and its three Replicas (p > 0.05). Of the 176 species observed in the basin, more than two-thirds showed anemochoric dispersal. The greatest floristic similarity was observed between the Downstream Area and *the* Central Area, *while the* Upstream *Area* had the highest number of exclusive species (94 spp.). There was no significant difference (p > 0.05) between the distribution of vascular epiphyte abundances in the different areas of the watershed. The richness of the upstream area was significantly different from the other two areas of the basin, but the downstream and central areas did not differ significantly from each other. The vertical distribution of epiphytes was different between the downstream and upstream areas, and between the central and upstream areas (p = 0.0001), but similar between the central and downstream areas (p > 0.05). The results reinforce the idea that the epiphytic community is dependent on atmospheric humidity. Among the main regulators of vascular epiphytic diversity in this study are the phytophysiognomic characteristics of the forest areas, macroclimatic factors (especially water seasonality) and factors that influence the microclimate of the fragments, such as the reduction of structural complexity or the creation of edges. In terms of conservation, the presence of 50 species under some degree of threat constitutes a strong argument for the conservation of forest fragments in the Sorocaba/Médio Tietê basin.

Keywords: Phytosociology of epiphytes, Atlantic Forest, Cerrado, Endangered species, Vertical distribution

CHAPTER 1

INTRODUCTION

Throughout the process of occupation and economic development of São Paulo's territory, there has been intense forest devastation, which has culminated in the reduction of countless plant formations to small, scattered fragments, mainly in the interior of the state. This situation has intensified, especially in the last decades of the 20th century, as a result of growing urban expansion and the cultivation of sugar cane, which occupies large areas of the state. The effects of the devastation of native forests can be seen both in the reduction of biodiversity and in the impairment of water quality, soil impoverishment and the intensification of the erosion process, among others.

The Sorocaba/Médio Tietê (SMT) river basin, located in the center-southeast of the state of São Paulo, covers an area of 11.827.824 km^2 , extending over 53 municipalities, 34 of which are located in its territory and 19 with rural portions in the basin area; it was defined as the "Water Resources Management Unit - UGRHI - Sorocaba/Médio Tietê" by Law No. 9.034/94, of 27/12/1994, which provides for the State Water Resources Plan for the 1994/95 biennium (ZERO REPORT, 2005).

In the context of the landscape, the Sorocaba/Médio Tietê basin is an extensive area, with remnants of natural vegetation that are completely fragmented. However, around 90% of the basin belongs to the Sorocaba administrative region (RELATÓRIO ZERO, 2005), which is the administrative region with the second largest natural vegetation cover in the state of São Paulo, behind only the coastal region (SMA, 2006).

Because it has different phytophysiognomies, the watershed is home to a number of plant species that are highly representative of the floristic richness of the state of São Paulo. The floristic diversity provided by the different plant formations that exist in the basin, in its upstream portions (Ombrophilous Forests), in the central part (Semideciduous Forests) and downstream (river mouth region - "ecological tension zones" between Ombrophilous Forests, Semideciduous Forests and Cerrados) enables the development of a very diverse fauna (ZERO REPORT,

2005). Despite this, the knowledge available on the remaining vegetation does not yet allow us to understand the mechanisms regulating biodiversity in these areas, nor to understand how recent changes have interfered in the processes of structuring and functioning of these forests.

The different phytophysiognomies that occur along the hydrographic basin have a significant influence on the composition of the vascular epiphytic community, given the peculiar environmental conditions characteristic of each type of forest, especially with regard to moisture supply. The Ombrophilous Forests are wetter, the Seasonal Forests have well-defined wet and dry periods or marked thermal variations, and the Cerrado, despite an average annual rainfall of between 1,200 and 1,800 mm, has a well-defined dry period; the regions known as "areas of ecological tension" have specific climatic characteristics. In addition, there is the forest's microclimatic gradient, which also influences the epiphytic vascular community. Characteristically, humidity increases from the canopy to the ground, while luminosity follows the opposite trend (BATAGHIN et al., 2008), the canopy being the part with the greatest thermal fluctuation (KERSTEN, 2006).

The role of biodiversity in the functioning of ecosystems has received increasing theoretical treatment (TILMAN, 1988; PIRES, 2001; GRIME, 2001; CALLAWAY et al., 2002; NAEEM, 2003; PIRES et al., 2005), however it is not yet clear how important biodiversity is in relation to the other component parts of the ecosystem, nor how much this degree of importance varies from one ecosystem to another (TILMAN; LEHMAN, 2001).

The main factors affecting the dynamics of forest fragments are: size, shape, degree of isolation, type of neighborhood and history of disturbances (PIRES, 1995; VIANA et al., 1992). These factors are related to biological phenomena that affect the birth and mortality of plants, such as the edge effect, genetic drift and interactions between plants and animals (VIANA; PINHEIRO, 1998). Among the most important consequences of the fragmentation process are the reduction of biological diversity, the disturbance of the hydrological regime of water basins, climate change, the degradation of natural resources and the deterioration of the quality of life of traditional populations (PIRES, 1995; VIANA, 1990).

Epiphytes are plants that establish themselves directly on the trunk, twigs, branches or leaves of trees, without the emission of haustorial structures, and the plants that support them are called phorophytes (BENZING, 1990). Because of their physiological and nutritional characteristics, epiphytes play a fundamental role in studies on anthropogenic interference in the environment, since they often absorb atmospheric moisture directly through their leaves or stems, making them more exposed to the action of pollutants (AGUIAR et al., 1981), as well as depending on tree and shrub vegetation for their establishment.

A striking feature of epiphytes, the so-called "vertical evolution", occurred in the exchange of spaces, i.e. in the search for more sunlight, the plants were exposed to conditions of greater stress for the acquisition of water and nutrients (BENZING, 1990); Thus, the canopy offers greater luminosity when compared to the understory (KIRA; YODA, 1989), but other resources are limiting in the canopy, such as the relative scarcity of nutrients, the instability of the substrate and, above all, water stress (LÜTTGE, 1989). Water stress is one of the greatest difficulties for survival above ground (LAUBE; ZOTZ, 2003). Much of the stratification can be attributed to the microclimatic variations that exist in forests (KERSTEN, 2006). Although these are ultimately determined by the macroclimate, the rhythms of exchange in forests are determined by the cycles established by the vegetation (PARKER, 1995).

Epiphytes also function as bioindicators of the successional stage of the forest, given that communities in secondary stages have lower epiphytic diversity than primary communities (MEIRA, 1997; BARTHLOTT et al., 2001). Epiphytes can reflect the degree of local preservation, since some groups are less tolerant of environmental variations resulting from deforestation and fires (SOTA, 1971).

According to Bonnet and Queiroz (2000), the density of individuals and the diversity of vascular epiphytic species is inversely related to the degree of alteration of the forest ecosystem, which allows epiphytes to be characterized as an important indicator of the state of conservation of forests. Wolf (2005), analyzing the epiphytic flora of areas with different levels of disturbance, concluded that forest disturbance has a negative effect on epiphytic biomass and its alpha diversity, as well as on the flora of the remaining trees. The fact that Conservation Units generally have a larger area and suffer less human action than unprotected forest fragments that are totally inserted in anthropogenic landscapes may contribute to the development of a richer and more diverse vascular epiphytic community in Conservation Units.

In addition, there are few systematic studies of the plant community in the Sorocaba/Médio Tietê basin, which does not allow detailed data to be presented on its floristic composition, nor quantitative data on its flora (ZERO REPORT, 2005). There is only one specific survey of vascular epiphytes in the study area, carried out by Bataghin (2009).

Kersten (2010) cites 10 published studies on vascular epiphytes in the state of Sao Paulo, of which only Bataghin et al. (2010) use epiphytes as indicators of the state of conservation of forests. Although the number of studies on epiphytes has recently increased in Brazil (KERSTEN, 2010), few studies focus on the role of vascular epiphytes in the dynamics of forest fragments and the importance of larger or smaller areas of forest, which reinforces the importance of this research. Furthermore, the Sorocaba/Médio Tietê River Basin Committee itself establishes as one of the main problems the lack of studies that can contribute to better characterizing, understanding and managing this river basin.

In this sense, the development of this thesis with vascular epiphytes in the Sorocaba/Médio

Tietê Hydrographic Basin contributes to the floristic characterization of this important component of diversity, to analyze the role of vascular epiphytes as environmental indicators, as well as to evaluate the ecological integrity of the forest fragments located in the basin, subsidizing conservation action plans in the area. The analysis and correlation of factors that govern the dynamics of forests are fundamental to understanding the mechanisms that maintain biodiversity, since the development of the community depends on the overall set of factors and not on their isolated action.

Vascular Epiphytes

Madison (1977) defines epiphytes as plants that, without being connected to the ground, use the support, but not the nutrients, of the forophytes on which they are supported at some stage in their lives. Epiphytism is defined by Bennet (1986) as a commensal interaction between plants, in which the dependent species (epiphyte) benefits only from the substrate provided by the host species (forophyte), taking nutrients directly from atmospheric humidity, without emitting haustorial structures. Kress (1986) and Wallace (1989) define epiphytes as plants that normally live above another and during any stage of their life cycle typically obtain all, or a significant part, of their water and mineral nutrients from sources other than the soil, without being parasitic.

Around 10% of all vascular species, approximately 25,000 species, make up the epiphytic community and are distributed in 84 families (KRESS, 1986). Gentry and Dodson (1987a) cite 83 families, 876 genera and around 29,000 epiphytic species worldwide. Kress (1986, 1989) lists 84 families, 879 genera and 23,466 epiphytic species distributed worldwide.

Epiphytes have spread to such an extent that few plant families have had great adaptive success. According to Madison (1977) and Benzing (1990), more than 95% of epiphytic species belong to the 20 richest families in this life form: Orchidaceae (68%), Araceae (4.6%), Bromeliaceae (3.9%), Piperaceae (2.5%), Ericaceae (2.3%), Melastomataceae (2.2%), Polypodiaceae (1.8%) and Cactaceae (0.5%).

Kersten (2006) highlights the following epiphytic families in Brazil, given their large number of species: Orchidaceae (35.9%), Bromeliaceae (18.7%), Polypodiaceae sensu lato (12.5%), Cactaceae (6.0%), Piperaceae (6.0%) and Araceae (4.1%). The same author also highlights the occurrence in Brazil of families such as Commelinaceae, Cyperaceae and Amaryllidaceae, although their numbers are less expressive. In addition, families such as Ericaceae, which are rich in epiphytes in other regions of the world, do not gain prominence in Brazil. Families such as Polypodiaceae and Bromeliaceae are very important in our territory (SMITH, 1962), and in many cases have the highest importance values (WAECHTER, 1992; GONÇALVES;

WAECHTER, 2002; KERSTEN; SILVA, 2002; GIONGO; WAECHTER, 2004).

Evolution and adaptive strategies

Although many epiphytic species are highly speciose, Benzing (1986) points out that the group has no distinctive characteristics and lacks unifying features. This can be partly attributed to the different origins: each family has developed specific habits separately. Another reason can be attributed to the diversity of habitats, especially in tropical forests, where humidity, irradiation and nutrients occur in numerous combinations.

The "vertical evolution" undergone by the epiphytic community is one of the last stages in the irradiation undergone by vascular plants, which began approximately 400 million years ago (KERSTEN, 2006). The exchange of space, in terms of sunlight, for conditions of greater stress for the acquisition of water and nutrients has been the defining characteristic of the group (BENZING, 1990), so the canopy offers greater luminosity when compared to the understory (KIRA; YODA, 1989), but other resources are limiting in the canopy, such as the relative scarcity of nutrients (little suspended soil), the instability of the substrate and, above all, water stress (LÜTTGE, 1989). Undoubtedly the most important abiotic factors for the growth of epiphytic flora are the acquisition and storage of water.

The fact that water stress is limiting for the community means that epiphytes are mainly found in wet forests; aridity excludes competitiveness for most species of vascular epiphytes (KERSTEN, 2006). For Zotz and Hietz (2001), the availability of nutrients and solar irradiance are generally less important for epiphytes, although they do influence the community.

The exposure of epiphytes to high levels of sunlight, fluctuations in temperature and humidity, as well as variations in the amount of water available (KIRA; YODA, 1989), makes their survival dependent on adaptations in both morphological and physiological aspects. Among the physiological aspects, those related to photosynthesis are vital. Most species with a CAM mechanism have an epiphytic habit (LÜTTGE, 2004) and around 57% of all epiphytes (possibly more than 15,000 species) use this mechanism; however, the C4 pathway has not yet been recorded for any species in this syntomy (ZOTS; HIETZ, 2001). Lüttge (2004) points out that CAM plants are typically adapted to arid environments and, although epiphytes are typical of tropical rainforests, they do not have direct or constant access to water. This metabolic pathway is the most suitable for accommodating species in the inconstant humidity observed in tree trunks (BENZING, 1990). The CAM mechanism is so important for epiphytes that Lüttge (2004) considers it a central element in the ecophysiology of this community.

Difficulty in accessing water is one of the greatest difficulties for survival above ground (LAUBE; ZOTZ, 2003). In forests, trees (forophytes) are characterized by mesomorphic leaves and C3 mechanisms, while epiphytes tend towards xeromorphism and have various mechanisms for absorbing and storing water (KERSTEN, 2006). Under unfavorable environmental conditions, such as lack of water and high temperatures, stomata often close to prevent dehydration, although water loss can persist through cuticular respiration. According to Helbsing et al. (2000), the first line of protection against desiccation is the cuticle, and the lowest cuticular permeability rates have been observed in epiphytic species.

Mineral nutrients can be captured directly from the atmosphere, either by suspended particles, by direct contact with rainwater or water leached from the canopy, or even by the accumulation of litter and animal waste deposited on trees. Another important source of nutrients is the accumulation of organic matter that epiphytes deposit on branches, forks or grooves in the bark of phorophytes. Nutrition can also come from animal sources (insectivorous plants) or plant sources, such as accumulated leaf litter (KERSTEN, 2006). In addition, most Neotropical terricolous species are infested with mycorrhizae and this interaction with mycorrhizae helps them to capture nutrients (RICHARDSON; CURRAH, 1995); the study of orchids, in particular, has shown different degrees of mycorrhizal association, from obligate to sporadic (LESICA; ANTIBUS, 1990).

Classification of vascular epiphytes

Vascular epiphytes can be classified on the basis of various factors, often the dependence on the forophyte, fidelity to the substrate and degree of exposure are the factors used. Benzing (1990) classifies epiphytes into two groups, holoepiphytes and hemiepiphytes, according to their fidelity to the substrate used.

1)   Holoepiphytes have an epiphytic habit throughout their life cycle and are subdivided into:

a)   Characteristic holoepiphytes: in a community they characteristically appear as epiphytes.

^ Obligate holoepiphytes: in a community they are never observed outside the epiphytic environment.

^ Preferred holoepiphytes: usually in a community they appear as epiphytes and can occasionally be found as terricolous.

b)   Facultative holoepiphytes: can grow in the same community both on trees and on the ground.

c) Accidental holoepiphytes: although they have no special adaptation for epiphytism, they occasionally grow to maturity on other plants.

2) Hemiepiphytes have a typically epiphytic habit for only part of their lives and are subdivided into:

a) Primary hemiepiphytes: germinate on the forophyte and later establish contact with the soil through geotropic roots.

^ Constrictors: can kill the supporting plant with their roots by preventing sap flow.

^ Non-constrictive: they never kill the forophyte, but only benefit from the support provided by them.

b) Secondary hemiepiphytes: germinate in the soil and later establish contact with the forophyte, losing their connection with the soil.

In addition to the above classification, epiphytes can also be classified according to the supply of resources, which takes into account the availability of water and nutrients throughout the year. When these resources are more or less stable, epiphytes are called continuous supply species. When these resources vary greatly, epiphytes are called pulse-supply species. According to Benzing (1990), in the same community, the presence of different microhabitats can lead to the occurrence of both types of species. Taking into account the water balance, the same author classifies epiphytes into two large groups, the poikilohydric and the homeoidric.

Poikiloids are species that can withstand great variations in humidity. During periods of drought they lose their color and take on a twisted appearance, but with increased humidity they return to their original form. They are often called "resurrection epiphytes" because of their ability to rehydrate even though they appear dead.

Homeophytes differ from poikilohydric plants in their great ability to slow down water loss and their low resistance to desiccation: (a) *Hygrophytes* - with thin leaves and delicate epidermis, they generally inhabit rainforests or humid environments, do not have xeromorphs and are perennial; desiccation, even for short periods, causes their death; (b) *Mesophytes* - shady species, restricted to the lower strata and common to humid places with a predominance of non-deciduous species, being more resistant to desiccation; and (c) *Xerophytes* - generally with narrow, long leaves and a thick epidermis; they are resistant to prolonged periods of water deficit.

Spatial distribution

The microclimatic variations that exist in forests are responsible for a large part of the vertical stratification that exists and, although these are ultimately determined by the macroclimate, the rhythms of exchange in forests are determined by the cycles established by the vegetation (PARKER, 1995). Important factors for the epiphytic flora, such as temperature, humidity, incidence of light, composition of the spectrum and polarization of the rays, vary differently in the forest (BENZING, 1995).

The temperature varies daily as you move away from the ground, with the canopy being the part with the greatest fluctuation in temperature range. Close to the ground, humidity remains practically constant and close to 100% for most of the day, while close to the canopy it can be between 50% and 60% (KIRA; YODA, 1989; LAUER, 1989; BENZING, 1995). The temperature itself can vary by several degrees between the canopy and the ground, directly influencing the relative humidity of the air.

The preference of epiphytic species for certain species of forophyte is associated by Brown (1990) with the moisture retention capacity, chemical composition and morphology of the bark. Although the moisture retention capacity of the bark may be indifferent for adult epiphytes, it has a strong influence on the establishment of young plants, for which small amounts of water are sufficient and essential for survival. The morphology of the bark (degree of roughness, periodic flaking, etc.) influences the establishment of the spores, the humidity and the amount of nutrients available (BENZING, 1995). Even factors such as wind direction and speed, as well as the shape and size of the seeds, can influence the number of epiphytic individuals (HERNADES-ROSAS, 2001).

Other factors, such as the degree of exposure (BENZING, 1990) or the architecture of the tree (SILLET, 1999), are important in the establishment and differentiated development of epiphytic species. The so-called edge effect can influence the development of epiphytes, either through luminosity, lower humidity or even greater incidence of wind, facilitating the transportation of spores, but also reducing humidity (BATAGHIN et al., 2008).

The occurrence of a positive and linear relationship between the size of the forophyte and the richness of epiphytes it supports is highlighted by Flores-Palacios and Gracio-Franco (2006), who state that there is a positive relationship both for certain species of forophytes and for the tree community as a whole. Although the dynamics of epiphyte populations is still a topic that is little considered in scientific studies, it is known that the density of individuals and species richness is inversely correlated to the degree of alteration of forest ecosystems (BONNET; QUEIROZ, 2000; BATAGHIN et al., 2010).

According to Callaway et al. (2001), the structure and diversity of epiphytic communities growing on different forbs can be influenced and even determined not only by the characteristics of the trees, but also by the interaction between epiphytic species. In fact, the microclimatic gradient and differences in substrate, which can be correlated both to the type of forest formation and to changes in the shape, angularity and diameters of the forbs, are environmental factors that determine the distribution of epiphytic flora.

Ecological importance

The importance of the structure of the epiphytic community for analyzing tropical diversity can be expressed both by the number of individuals and their relative abundance (BIERREGAARD et al., 1992). The epiphyte community contributes to increasing the structural complexity of tropical forests, occupying everything from the canopy to the soil (FONTOURA, 2001), as well as positively influencing ecosystem processes and maintenance (LUGO; SCATENA, 1992).

Epiphytes are an important source of resources for animals living in the canopy, providing food, water or even material for building breeding sites (NADKARNI, 1984). For Hadel (1989), pond epiphytes form phytotelm environments and are essential for algae, numerous invertebrates and various vertebrates (small amphibians and reptiles) that use or depend on these still water deposits to live or complete their life cycles. Lugo and Scatena (1992) point out that the plant mass of epiphytes is of great importance in the cycling of nutrients and water within forests.

The diversity and abundance of vascular epiphytes can vary depending on the substrate, humidity and shade provided by the tree species in the occupied communities, which makes them suitable for use as indicators of the state of conservation of ecosystems (TRIANA-MORENO et al., 2003). Wolf (2005), studying the epiphytic flora in forests with different levels of disturbance, concluded that forest disturbance has a negative effect on epiphytic biomass and its alpha diversity, as well as on the flora of the remaining trees. Barthlott et al. (2001) observed a decline in species richness and a reduction in the number of species as a result of an increase in the degree of human interference. However, the epiphytic flora has shown itself to be resistant to disturbances when logging spares large trees, highlighting the importance of larger forophytes, which are essential for epiphytes that need suspended soil and can also serve as a source of seeds for young trees (BATAGHIN et al., 2010; DETTKE et al., 2008). The changes brought about by human activity are forcing a shift from mesic species, which are common in humid places (but not restricted to them), to poikiloid species, which are resistant to large variations in humidity.

The nutritional dependence of epiphytes on environmental conditions to obtain water and nutrients from the air allows them to be used for biomonitoring atmospheric conditions and, in particular, anthropogenic pollution. Nimis et al. (1990) and Henderson (1993) point out that mainly lichens have been used for this purpose. However, other epiphytic species have been shown to be applicable to environmental monitoring (GRACINO et al., 2003; WANNAZ; PIGNATA, 2006; MANETTI et al., 2009).

Epiphytes are responsible for increasing genetic diversity and promoting the (re)distribution of resources in tree trunks. The organic matter accumulated by epiphytes provides a rich source of nutrients for fauna and above-ground vegetation (NADKARNI, 1981; INGRAM; NADKARNI, 1993). They contribute to the retention of water directly from the mist (CLARK et al., 1998), and also help with biological activities in the treetops, including nitrogen fixation, keeping the environment humid, either through the evaporation of water stored in the biomass or through evapotranspiration (WEAVER, 1972). Thus, epiphytes are an important source of moisture and nutrients in forests, especially during the dry seasons.

The epiphytic flora contributes to the diversification of niches and microhabitats, considerably increasing the physical space and food available, as well as serving as a reproductive refuge for many species of animals (BENZING, 1986).

Geographical distribution

The distribution of epiphytes is quite uneven around the tropics (Kersten 2006), Africa is considerably poorer in species than the Americas, with Asia being an intermediate region (MADISON, 1977). The lowest specific richness of epiphytes seems to occur in Oceania; according to Wallace (1989), only 350 species are found in the whole of Australia. Attempts to explain the great diversity in the tropics are usually based on paleoclimatic fluctuations and, consequently, on the formation of forest refuges, such as the shrinking of forests into "capes" in which isolation led to speciation, given the reduction in global temperature and the expansion and shrinking of dry and wet vegetation (KERSTEN, 2006). The separation of the great continent of Gondwana (around 120 million years ago) and the formation of the Andes chain are factors that deserve to be highlighted (GENTRY, 1982).

In the Neotropics, the distribution of epiphytes is also irregular. On the Yucatán peninsula (Mexico), there are only 107 species (OLMSTED; JUAREZ 1996). In Brazil, several isolated studies have shown a greater number of species: in Rio de Janeiro (FONTOURA et al., 1997) and Sao Paulo (MAMEDE et al., 2001), 300 and 160 species were found, respectively. Barros et al. (2002), based on studies carried out in the Parque Estadual das Fontes do Ipiranga in the city of Sao Paulo and the current study, recorded 145 epiphytic species, without considering the pteridophytes occurring in the area.

According to Smith (1962), there is a general ecological trend towards a greater number of plant species in the tropics and a decrease in richness towards the poles. Similarly, Waechter (1998) points out that the abundance and richness of epiphyte flora decreases rapidly after 30° south latitude, the limit of influence of the tropical masses.

The attempt to obtain hydric resources (from the atmosphere) and to overcome hydric stress makes the wet regions or forests of the globe the main centers of diversity for epiphytic flora. The greater epiphytic diversity observed in wet montane forests in the American tropics seems to follow a general trend shown by angiosperms (GENTRY; DODSON, 1987a).

Moran (1995), studying mountainous regions, associated the richness of pteridophyte species with the varied microenvironments created by the different elevations, slopes, luminosity, soils, rock types and microclimates. According to Benzing (1990), ombrophilous forests generally have epiphytes and in addition to this, Schütz-Gatti (2000) and Kersten and Silva (2001) report that in these formations the diversity can be such that dozens of species can occur in a single tree (phorophyte).

In desert locations, epiphytes are less common and have a lower number of taxa, but not necessarily a lower abundance (KERSTEN, 2006). Some cactus and shrub forests can support dense communities of bromeliads and orchids (BENZING, 1990). Kersten (2006) mentions that, even in an inhospitable situation with regard to humidity, more than 200 individuals of *Tillandsia stricta* (Bromeliaceae) were observed on individuals of *Lagerstroemia indica* (Lythraceae) used in urban afforestation (Curitiba, PR). Even in extremely arid climates, where temperatures can vary from 40 °C during the day to 15 °C at night, or in high mountains subject to freezing and snow, some types of bromeliad can grow (ROUSSE, 1994). The occurrence of epiphytes in areas with well-defined dry seasons (due to rain, freezing, etc.) can be considered an indication of a short water deficit (SCHIMPER, 1888). Gentry and Dodson (1987b), studying three forests subject to different degrees of humidity, reported that the reduction in the number of epiphytic species caused by the reduction in the unit is significantly greater than that observed for the arboreal, shrub or herbaceous floras.

The reason for the increase in diversity in rainforests can be attributed to the ability of epiphytes to achieve a more elaborate partitioning of niches occurring in different regions of the forophyte, which contributes to the high alpha diversity. Another cause of this increase may be related to the variety of forest formations characteristic of mountainous regions, where these impose genetic barriers that increase beta diversity (GENTRY; DODSON, 1987b; NIEDER et al., 1999).

Characterization of the Phytophysiognomies of the Sorocaba/Médio Tietê river basin

Cerrado

The Cerrado covers approximately 25% of Brazil's territory, around two million square kilometers, and is the country's second largest biome in terms of area. Located on the Central Plateau, it occurs at altitudes ranging from around 300 to over 1,600 meters. The Cerrado extends across the states of Goiàs, Tocantins and the Federal District, as well as parts of Bahia, Cearà, Maranhao, Mato Grosso, Mato Grosso do Sul, Minas Gerais, Piaui, Rondônia and Sao Paulo, and also occurs in isolated areas to the north in the states of Amapà, Amazonas, Parà and Roraima, and to the south on small islands in Paranà (RIBEIRO; WALTER, 1998), even reaching neighboring countries such as Bolivia and Paraguay (TOPPA, 2004).

One of the main characteristics of cerrado regions is the presence of dry winters and rainy summers. The occurrence of two well-defined seasons characterizes the concentrated distribution of rainfall throughout the region, with a direct influence on the vegetation. The rainy season runs from October to March and the dry season from April to September (RIBEIRO; WALTER, 1998). The climate also has a temporal influence on the origin of this vegetation, as rainfall over geological time has weathered the soils, leaving them poor in essential nutrients. The Cerrado's main climate classification, according to Koppen, is Aw - tropical rainy. The Aw climate coincides with the distribution of most of the world's savannas (RICHARDS, 1996). The average annual rainfall is 1,500 mm, ranging from 750 to 2,000 mm (TOPPA, 2004).

Located on the southern edge of the Cerrado-dominated region, the state of Sao Paulo originally had patches of vegetation covering around 14% of the state (KRONKA et al., 1998), dispersed in a predominantly forested landscape. Based on the Vegetation Map of Brazil, the areas covered by cerrado vegetation, plus the so-called ecological tension zones, i.e. the transition between the cerrado and neighboring forest formations, corresponded to approximately 30% of the state's surface (DURIGAN et al., 2006).

According to Silva and Bates (2002), the Cerrado is recognized as one of the largest, richest and possibly most threatened ecosystems in the world. The rapid destruction to which it has been subjected and the high biological richness of the Cerrado have placed it among the priority biomes for biodiversity conservation at a global level (MYERS et al., 2000). However, in recent decades, areas of Cerrado have been drastically reduced throughout Brazil, with only less than 34% of its original extent remaining (MACHADO et al., 2004).

Semideciduous Seasonal Forest

The Semideciduous Seasonal Forest can also be called inland rainforest (MAACK, 1968). The original area extended from Rio Grande do Sul, north of the Rio Jacui, to Minas Gerais, always below the altitudinal limit of the araucaria. In Santa Catarina and Rio Grande do Sul, this type of forest comprised large areas of the Uruguay River basin. In Paranà, it was present in around 1/3 of the state's total area. In the state of Sao Paulo, it occurred throughout the Paranà basin and in the entire area not covered by savannahs. It is limited to the Serra do Mar, in the states of Sao Paulo, Minas Gerais and Rio de Janeiro, and to the Serra Geral, in the south, often entering the river valleys, invading the adjacent forests. To the south, these forests advance deep into the territory of Paraguay and Argentina (HUECK, 1972).

The climate with two well-defined seasons, one rainy and the other dry, or a marked temperature variation (winter and summer) is the main reason for the occurrence of the Semideciduous Seasonal Forest. These peculiarities of the climate are determinants of leaf deciduousness (mainly of arboreal individuals) as a response to the period of water deficiency, or to the reduction in temperature in the cold months (VELOSO et al., 1991).

This type of forest formation can have trees that reach 30 to 40 meters in height (when in flatter regions with more developed soils), although it does not form a continuous top cover. The characteristic that determines its classification as semi-deciduous is the loss of leaves during the winter, which occurs in around 40% to 50% of the trees in this formation. Families such as Lauraceae, Meliaceae, Fabaceae and Rutaceae are responsible for the second tree layer, which is very dense and evergreen, formed mainly by trees (KERSTEN, 2006). The understorey is usually made up of Euphorbiaceae, Moraceae and Rubiaceae trees (VELOSO et al., 1991).

Aspidosperma polyneuron, Gallesia integrifolia, Astronium graveolens and *Parapiptadenia rigida* stand out among the tallest trees in the forest. Tree species predominate in the canopy and forest understorey of remnants of Semideciduous Seasonal Forest (SOARES-SILVA; BARROSO, 1992; SOARES-SILVA et al., 1992). According to Kersten (2006), in the forest canopy, which is more or less continuous, the Lauraceae families stand out, notably *Nectandra megapotamica,* Meliaceae (*Cabralea canjerana, Trichilia* spp.) and Fabaceae (*Lonchocarpus* spp. and *Machaerium spp.*). In the forest understory, in addition to many young individuals of species from the upper strata, *Sorocea bonplandii, Actinostemon concolor* and *Euterpe edulis* stand out, as well as some species of Meliaceae (*Trichilia* and *Guarea*) and Rutaceae *(Esembeckia).* In the herb layer, the families of epiphytes and lianas are basically the same as those mentioned for the Araucaria forest, although there are variations between the most frequent species.

Dense Ombrophilous Forest

Dense Ombrophilous Forest is a name established by Ellemberg and Mueller-Dombois (1965/1966) as a reference to the forest's water affinity. These authors first used the terms "Dense" and "Open" to divide the forest within the intertropical space (VELOSO et al., 1992). However, this formation is also known by the term "Tropical Rain Forest" coined by Schimper (1903) and refined by Richards (1952).

This vegetation is characterized by the presence of phanerophytes, as well as abundant woody lianas and epiphytes (VELOSO et al., 1992). However, the main ecological characteristics of this type of forest are observed in ombrophilous environments, which are dictated by their ombrothermal characteristics. The Ombrophilous Dense Forest is conditioned by the tropical climatic factors of high temperatures (average of 25 °C) and high rainfall, well distributed throughout the year (no biologically dry period). In addition, dystrophic and, exceptionally, eutrophic latosols, which originate from various types of rock, predominate in this type of forest (VELOSO et al., 1992).

IBGE (2012), presents a subdivision for Ombrophilous Dense Forests based on their distribution:

Alluvial Formation - topographically similar, it always presents repetitive environments, within the alluvial terraces of the pluvials.

Lowland Formation - situated between 5 m and 100 m above sea level between latitudes 4° N and 16° S; from 5 m to 50 m above sea level between latitudes 16° and 24° S; and from 5 m to 30 m between latitudes 24° and 32° S.

Submontane Formation - located on the slopes of plateaus and/or mountain ranges. It occurs between 100 m and 600 m, between latitude 4° N and latitude 16° S; from 50 m to 500 m between latitude 16° and 24° S; and from 30 to 400 m between latitude 24° and 32° S.

Montane Formation - located high up on the plateaus and/or between 600 m and 2000 m above sea level, for the range between 4° N and 16° S latitude; from 500 to 1500 m between 16° and 24° S latitude; and between 400 and 1000 m above sea level between 24° and 32° S latitude.

Alto-Montana Formation - situated above the limits established for the Montana Formation.

In general, the Atlantic Dense Ombrophilous Forest physiognomies can be divided into two large groups, according to their geological basis and their position in the landscape. The first group occurs on Cenozoic sedimentary plains and comprises the Alluvial, Lowland and Submontane formations, the latter on colluvial deposits. The second group occurs on mountain ranges or folding systems underlain by Precambrian rocks, comprising the Submontane, Montane and High Montane formations (IBGE, 1992; LEITE, 2002; RODERJAN et al., 2002).

The High-Montane Ombrophilous Dense Forest, which is part of the vegetation recorded in the Montante Area studied, is a mesophanerophilous tree formation of approximately 20 meters in height, usually located on the tops of mountains with lithic soils or with peaty accumulations when recorded in depressions. Its structure is made up of phanerophytes with thin trunks and branches, small, leathery leaves and thick bark with fissures (IBGE, 2012), which favors epiphytism. The flora is represented by globally dispersed families, although their species are often endemic (VELOSO et al., 1992).

The Dense Ombrophilous Forest is the predominant plant physiognomy in the coastal region of southern Brazil (IBGE 1992). Climatic factors, especially abundant humidity and high temperatures, provide the exuberance of the vegetation, not only in terms of the size of the individuals or the speed of their development, but also in terms of the high species richness. Even when it is in an extratropical zone in the south of Brazil, it has an essentially tropical physiognomy, resulting only in the absence of some typical species and, on the other hand, the potential for endemism (BLUM et al., 2011). This environmental diversity, the result of the interaction of multiple abiotic factors, is an important aspect of this phytoecological region, with considerable influence on the dispersal of species and the structural development of the forest, resulting in distinct physiognomies (WETTSTEIN, 1970; LEITE; KLEIN, 1990).

Ecotone areas

According to Odum (1988), an ecotone is a transition between two or more different communities in which many of the organisms from each of the overlapping communities are present, as well as organisms characteristic of the ecotone which are often restricted to it.

An ecotone is an area of ecological tension in which one type of vegetation gradually replaces another (WALTER, 1986). Generally, the occurrence of two types of vegetation under the same general climatic conditions takes place in competition. The result of this competition will depend on the microclimatic conditions, as well as the conditions, type and texture of the soil and the relief of the site, so there may be a diffuse "mixture" or a kind of mosaic of the two types of vegetation.

In the area of the Sorocaba/Médio Tietê river basin, it is possible to identify the presence of two types of ecotone: one between the Dense Ombrophilous Forest and the Semideciduous Seasonal Forest in the upstream area and the other between the Semideciduous Seasonal Forest and the Cerrado in the downstream area of the river basin.

The ecotone of the Montante Area of the Sorocaba Médio Tietê Basin is made up of the contact between the Semideciduous Seasonal Forest and the second physiognomic group of the Ombrophilous Dense Forest classification, mainly the Ombrophilous Dense Montana and Alto-Montana formations. This peculiar vegetational characteristic creates a unique environment for the epiphytic vascular component, allowing species to occur in both physiognomies, although in certain forest fragments the availability of resources and interspecific competition can act on the community in general.

The border between the Semideciduous Seasonal Forest and the Cerrado is responsible for forming the ecotone vegetation of the Downstream Area of the Sorocaba Médio Tietê Basin. In this area, as reported by (ULHMANN, 1997), the invasion of the forest over the cerrado is notorious, supposing that the forest interferes with the fertility patterns of the soils and that some other characteristic of an unknown nature is determining the implantation of forests in those places. According to Linsingen et al. (2006), the most frequent tree species in the ecotone zone between the Semideciduous Seasonal Forest and the Cerrado are: Pera obovata (Klotzsch) Baill, Ocotea corymbosa (Meisn.) Mez, Laplacea fruticosa (Schrad.) Kobuski, Myrcia breviramis (O. Berg) D. Legrand, Vochysia tucanorum Mart, Copaifera langsdorffii Desf., Couepia grandiflora (Mart. & Zucc.) Benth. ex Hook. f., Qualea cordata (Mart.) Spreng., Serjania gracilis Radlk. and Schefflera vinosa (Cham. & Schltdl.) Frodin & Fiaschi (LINSINGEN et al., 2006).

CHAPTER 2

OBJECTIVES AND HYPOTHESES

The objectives of this research were: (a) to evaluate the floristic composition of vascular epiphytic vegetation along the Sorocaba/Médio Tietê river basin; (b) to quantitatively evaluate the epiphytic component in selected areas of the Sorocaba/Médio Tietê river basin; (c) to analyze the distribution of epiphytic species along the specified basin; (d) to analyze the vertical distribution of epiphytes in the study area; (e) evaluate the floristic composition of the vascular epiphytes present at the sites in comparison with studies carried out in areas of the same phytophysiognomic type; (f) analyze the epiphytic flora associated with the different phytophysiognomies in the study area; (g) analyze the state of conservation of the forests along the Sorocaba/Médio Tietê basin based on the vascular epiphytic component. The following hypotheses were tested:

• Considering the different phytophysiognomies in each of the basin's portions (upstream, in the central part and downstream), is there a difference in the floristic composition of vascular epiphytes between these portions?

Null hypothesis - There is no difference in floristic composition between these different portions of the watershed;

• Considering the possible differences in macroclimatic and microclimatic gradients in the different portions of the basin, is there a difference in the vertical distribution of vascular epiphytes upstream, in the central part and downstream of the basin?

Null hypothesis - The vertical distribution of epiphytes does not differ between the portions of the watershed;

• Taking into account that Conservation Units suffer less anthropogenic action than other forest fragments, does the floristic composition of vascular epiphytes in the core sites (areas protected as Conservation Units) differ from their replicas (forest fragments not protected as Conservation Units)?

Null hypothesis - There is no difference in the floristic composition of the core sites (Conservation Units) and their replicas (areas not protected as Conservation Units);

• Considering the possible microclimatic variations in protected and unprotected areas, does the vertical distribution of vascular epiphytes in the core sites differ from their replicas?

Null hypothesis - There is no difference in the vertical distribution of epiphytes between the core sites and their replicas;

- Is there a correlation between the epiphytic species found at the sites and the epiphytic species found in studies carried out in areas of the same phytophysiognomic type?

Null hypothesis - There is no correlation between epiphytic species between the different surveys.

CHAPTER 3

MATERIAL AND METHODS

Study Area

The Sorocaba/Médio Tietê basin (Figure 1) is located in the center-southeast of the state of São Paulo, between the coordinates 22°30' to 23°45' S, and 48°15' to 47°00' W. Due to its size and intra-regional peculiarities, it is subdivided into six sub-basins: Lower Middle Tietê (located in the extreme northwest of the basin), Middle Tietê (located in the center-north of the basin), Lower Sorocaba (located in the south/southeast of the basin), Middle Sorocaba (located in the center-south portion of the basin), Upper Middle Tietê (located in the eastern portion of the basin) and Upper Sorocaba (located in the extreme southeast of the basin) (ZERO REPORT, 2005).

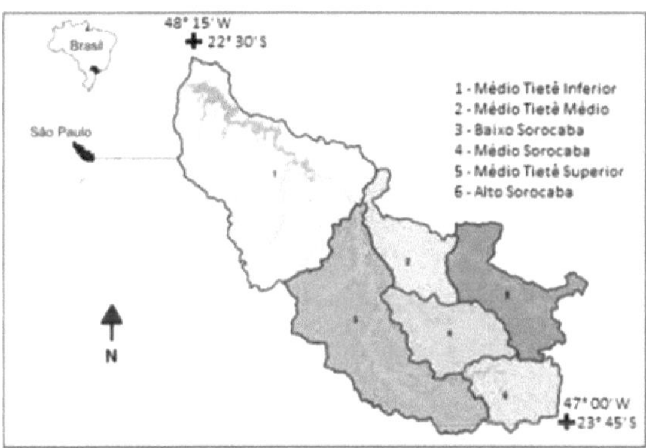

Figura 1:   Location of the Sorocaba/Médio Tietê river basin and its sub-basins.

The Sorocaba/Médio Tietê basin receives water from the Alto Tietê basin (to the southeast), and has the Tietê/Jacaré basin downstream (to the northwest). The Piracicaba, Capivari and Jundiai river basins, which make up the Piracicaba/Capivari/Jundiai basin, are the northeastern boundary, while to the south-southwest-northwest are the Upper and Middle Paranapanema basins and in the extreme south-southeast there is a small interface with the Ribeira de Iguape and Litoral Sul basins (ZERO REPORT, 2005).

The Brazilian Soil Classification System, defined by Embrapa (1999) mapped 38 types of soils present in the Sorocaba/Médio Tietê basin, organized into eight classes: Red-Yellow Argisols (15 soil types); Red-Yellow Latosols (eight types); Red Latosols (five types); Yellow Latosols (one type); Red Nitosols (three types); Neosols (four types); Planossolos Hàplicos (one type) and Cambissolos (one soil type).

The Tropic of Capricorn passes through the basin, characterizing it as a transition zone from tropical to temperate. According to the Koppen system, the Sorocaba/Médio Tietê basin has climatic conditions of the types Cwa - warm subtropical, with drier winters (this is predominant in the basin area), Aw - rainy tropical with dry winters (less frequent in the Central and Downstream Areas), Cfa - hot subtropical, constantly humid, with less dry winters (near the municipalities of Ibiùna and Piedade) and Cfb - humid subtropical, without a defined dry season (near the municipality of Sao Roque) (SETZER, 1966).

The Sorocaba/Médio Tietê river basin is part of the Atlantic Forest domain, comprising forest formations such as ombrophilous and seasonal forests, as well as savannahs and regions known as "ecological tension zones" - ecotones between ombrophilous and seasonal forests, and between these and savannahs. According to the ZERO REPORT (2005), nine types of vegetation can be found: Cerrado stricto sensu and cerradoes; Semideciduous Seasonal Forest; Dense Ombrophilous Forest; Riparian or Riparian Woodlands; Capoeira; Wetlands/Variegated Areas; Reforestation; Annual (Temporary) Crops and Anthropic Fields (Pastures).

Although the study area has 17 protected areas, only 12.09% of the watershed has natural forest cover (including original forests and the different successional stages) and 87.91% is occupied by anthropogenic or human-influenced cover, particularly pastures (67.64%) (CBH-SMT; FABH-SMT, 2008). The relief of the flattened hills of the middle and upper middle Tietê sub-basins has favored the occupation of the territory by agricultural activities in these basins. As a result, some municipalities have less than 2% of their area occupied by native vegetation, except for the upper part of the upper middle Tietê sub-basin, especially in the region of the municipalities of Sao Roque and Cabreùva, where the vegetation is more expressive. The lower middle Tietê, middle Sorocaba and lower Sorocaba sub-basins, despite not having the percentage of vegetation required by legislation, have significant forest remnants, totaling around 24,000 ha of forest formations for these three sub-basins. The Alto Sorocaba sub-basin, the region where the Sorocaba River rises, is the only one with the percentage of vegetation required by law. In this sub-basin, some municipalities have more than 50% of the total area with some form of vegetation (ZERO REPORT 2005).

Floristic study of vascular epiphytes

Given the extent of the river basin, almost 12,000 km^2 , and its location with different phytophysiognomic types, the study sites were established by dividing the river basin into three large areas according to vegetation type (Figure 2): Downstream Area (areas of Cerrado and ecotone between Cerrado and Semideciduous Seasonal Forest), Central Area

(Semideciduous Seasonal Forest) and Upstream Area (predominantly Dense Ombrophilous Forest and ecotone areas between Dense Ombrophilous Forest and Semideciduous Seasonal Forest).

For the qualitative analysis of the vascular epiphytic individuals, the methodology proposed by Kersten and Silva (2002) was used. Monthly trips were made to the research areas over a period of one year, and all the fertile species were collected and herborized using a methodology guided only by a compass, taking walks for around 30 hours (total floristic sampling period) at each site. In addition to the floristic inventory of vascular epiphytes carried out in the 12 quantitative sites (called core sites and replicas), a floristic survey was also carried out in a further nine sites (called qualitative sites) in areas not previously covered, thus sampling 21 sites (forest fragments) in the floristic study, and in order to explore different environments in the hydrographic basin. The sterile individuals were marked with plaques and geographic coordinates for later collection and/or cultivation until flowering, and recorded as reference specimens.

The herborization of the vascular epiphytes followed the usual procedures for floristic surveys (FIDALGO; BONONI, 1989) and the corresponding exsiccates were added to the UFSCar - Sao Carlos Herbarium. The collected material was identified using specialized literature, by comparison with material already deposited in herbaria in the region or by consulting specialists. The validity and spelling of the species names was checked on the specialized websites: w3Tropicos (www.tropicos.org), ePIC (www.rbgkew.org.uk/epic) and World Checklist of Selected Plant Families (http://apps.kew.org/wcsp/home.do), using the abbreviations of the authors suggested by Brummitt and Powell (1992).

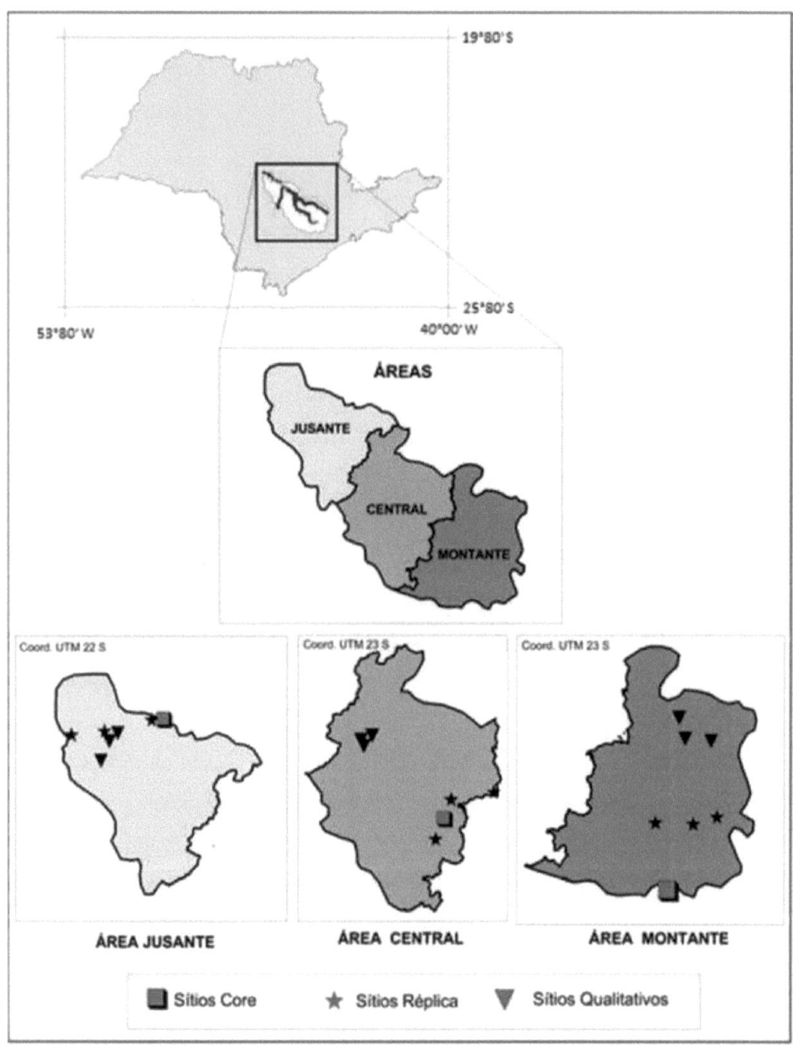

Figura 2: Location of the three large areas of the Sorocaba/Médio Tietê river basin and distribution of the sampling sites.

The species surveyed were classified into ecological categories (Table 1), according to their relationship with the forophyte, following the classification proposed by Benzing (1990). They were also classified according to their dispersal syndrome into two categories: anemochoric (wind dispersal) and zoochoric (animal dispersal).

Table 1 - Classification of species into ecological categories according to their relationship with the forophyte.

Abbreviation	Ecological category	Description
HLC	Characteristic holoepiphyte	They usually grow on top of other plants.
HLF	Facultative holoepiphyte	In the same community, they can grow both as epiphytes and as terricoles.
HLA	Accidental holoepiphyte	Generally terricolous, but can occasionally develop as epiphytes.
HMP	Primary hemiepiphyte	They begin life as epiphytes and later establish contact with the soil.
HMS	Secondary hemiepiphyte	They germinate in the soil and, when they come into contact with a forophyte, they degenerate the basal portion of the root/cauline system, becoming epiphytes.

Source: adapted from Benzing (1990).

To explore the similarity between the different areas (upstream, central and downstream) and between the different sites in the study area, based on the taxa identified, a cluster analysis was used, using the UPGMA method and Jaccard's similarity coefficient (MAGURRAN, 1988; ODUM, 1988).

In order to analyze the floristic composition of the vascular epiphytes at the sites in the basin in relation to other sites sampled of the same phytophysiognomic type, a compilation of lists from other published works was organized using Microsoft Excel. To explore the similarity between the different studies carried out, in comparison with the vascular epiphytic community of the Sorocaba/Médio Tietê basin, based on the taxa identified, a multivariate cluster analysis was used, applying Jaccard's similarity coefficient to a species presence-absence table, and UPGMA as the grouping method. The graphical form used to represent the final result of the grouping is a dendrogram.

Determining the sites for the quantitative study

The determination of the sites for the quantitative study of the vascular epiphytic community was based on the mapping of the watershed using the MapInfo version 8.5 and IDRISI Andes programs, and the IBGE planialtimetric maps at the 1:250,000 and 1:50,000 scales, thus identifying the areas where the sites would be located upstream, in the central part and downstream of the watershed. Preferably, areas were selected in which the "core sites" were located within Conservation Units and their replicas were located in surrounding forest fragments, especially on private properties, so that the type of neighbourhood of these fragments was characterized by anthropic use. The nomenclature adopted of Core Sites (Conservation Units), Replicate Sites (qualitative and quantitative surveys outside Conservation Units) and Qualitative Sites (only qualitative surveys outside Conservation Units) serves to better identify the forest fragments studied, however, it does not mean that, for example, the Core Site is an area in a better state of conservation, although this would be expected as they are Conservation Units. To help determine the sampling sites, geographic coordinates were taken with the aid of a GPS, so that the data obtained could be spatialized in the LAPA/UFSCar georeferenced database.

Selection of Forophytes

At each selected site, 90 forophytes (individual trees) with a DBH (diameter at breast height) ≥ 20 cm were sampled, since this inclusion criterion meant that all the forophytes reached the canopy of the forest fragments studied. Each forophyte was marked and numbered with identification plates. For the tillered individuals, those with at least one tiller within the inclusion criteria were included. The Quadrant Point method (COTTAN; CURTIS, 1956) was used to select the phorophytes, where points are established that act as the center of a Cartesian plane, defining four quadrants at each point, where the arboreal individual (phorophyte) that was closest to the central point within each quadrant was sampled.

Phytosociological study of vascular epiphytes

For the quantitative study, the forophytes selected within the inclusion criteria were considered. The forophytes were divided according to the method proposed by Johansson (1974), Braun-Blanquet (1979) and Steege and Cornelissen (1989), into six strata for the purpose of analyzing the vertical distribution of epiphytes: low stem (up to 1.3 m from the ground), medium stem, high stem, crown base, inner crown and outer crown, in which all the vascular epiphytic species were recorded. Each epiphytic species occurring in these strata was given a score for its respective abundance according to Kersten and Silva (2002), namely: 1 - one or a few isolated individuals; 2 - more extensive groupings or several isolated individuals; 3 - abundant, in many cases forming an almost continuous cover over the forophyte.

The absolute frequencies of occurrence in the strata (FAr) and in the forophytic individuals (FAi) were calculated, as well as the epiphytic importance value (VIE), obtained from the scores given to the epiphytes. The formulas used for these analyses, based on Kersten and Silva (2002), are as follows: $FAr = (nr\ na^{-1}).100$; $FAi = (ni\ nt^{-1}).100$; $VIE = [vt.(\sum vt)^{-1}].100$, where nr = number of regions (strata) where the epiphytic species occurs; na = total number of regions (strata) sampled; ni = number of individuals where the epiphytic species occurs; nt = total number of forophytic individuals; vt = sum of the scores obtained by the species. The average score given to the species in the intervals (forophytic strata) was also calculated in an attempt to express the average abundance on the forophytes. In this way, a species with an average score above 2 occupies large areas of the forophyte, while another with an average score close to or equal to 1 is not very expressive on the trees. Another parameter used was the sum of the scores of the epiphytic species, called the absolute value (AV), in each stratum or forophyte considered, which represents an estimate of abundance and richness.

Shannon's diversity index (H') was calculated based on the data on the occurrence of species on the individual forophytes. The similarity of the epiphytic vascular component, based on the presence or absence of total species and the vertical distribution in the forophyte, in relation to the environments studied, was calculated using cluster analysis using Jaccard's coefficient, with UPGMA (MAGURRAN, 1988) being used as the grouping method.

The data obtained in the field regarding the vascular epiphytes was organized in Microsoft Excel spreadsheets and then analyzed. In order to verify possible correlations between the floristic composition, the vertical distribution of the vascular epiphytes and the sites analyzed, a multivariate cluster analysis was applied, based on the presence/absence of species, and VA values, to verify the similarity between the sites and areas sampled. The Mann-Whitney test was used to test for possible variations in the vertical distribution of the epiphytic community. Possible differences between sites were verified using the t-test. The computer programs XLSTAT version 5.2, Past. version 2.17b (HAMMER et al., 2001), BioEstat 5.0 (AYRES et al., 2007) and MVSP version 3.0 (KOVACH, 1993) were used to carry out the statistical analyses.

CHAPTER 4

RESULTS AND DISCUSSION

UPSTREAM AREA

Floristic survey of vascular epiphytes in the downstream area of the Sorocaba/Médio Tietê river basin

In the floristic survey carried out in the Downstream Area of the Sorocaba/Médio Tietê river basin, characterized as an ecotone region between the Semideciduous Seasonal Forest and Cerrado vegetation types, 56 species were found, belonging to 28 genera and eight families (Table 2). The Shannon index for the epiphytic vascular community in the downstream area was H'= 2.948, the equability J = 0.732 and the Margalef richness index (d) was 6.470.

Table 2 - List of vascular epiphyte species found in the phytosociological survey of the Downstream Area of the Sorocaba Mèdio Tietê river basin and their respective Ecological Categories (EC) - HLC: Characteristic Holoepiphyte; HLF: Facultative Holoepiphyte; HLA: Accidental Holoepiphyte; HMP: Primary Hemiepiphyte and HMS: Secondary Hemiepiphyte. Dispersal form (Disp.) - Zo: Zoochoric; An: Anemochoric. Reg: HUFSCar herbarium registration number (Im: Digital image).

Family	Species	EC	Disp.	Reg.
ARACEAE				
1	Philodendron appendiculatum Nadruz & S.J. Mayo	HMS	Zo	8487
2	Philodendron bipinnatifidum Schott	HMP	Zo	Im
BROMELIACEAE				
3	Acanthostachys strobilacea (Schult. f.) Klotzsch	HLC	Zo	8431
4	Aechmea apocalyptica Reitz	HLF	Zo	8455
5	Aechmea bromeliifolia (Rudge) Baker	HLC	Zo	8432
6	Aechmea distichantha Lem.	HLF	Zo	8503
7	Billbergia amoena (Lodd.) Lindl.	HLC	Zo	8379
8	Tillandsia funckiana Baker	HLC	An	8458
9	Tillandsia recurvata (L.) L.	HLC	An	8429
10	Tillandsia stricta Sol. ex Sims	HLC	An	8428
11	Tillandsia tricholepis Baker	HLC	An	8430
12	Tillandsia usneoides (L.) L.	HLC	An	8460
13	Vriesea bituminosa Wawra	HLC	An	8515

Table 2 - Contents...

Family	Species	EC	Disp.	Reg.
BROMELIACEAE				

14	*Vriesea procera* (Mart. ex Schult. & Schult.f.) Wittm.	HLC	An	8508
CACTACEAE				
15	*Cereus alacriportanus* Pfeiff.	HLC	Zo	Im
16	*Epiphyllum phyllanthus* (L.) Haw	HLC	Zo	8474
17	*Lepismium cruciforme* (Vell.) Miq.	HLC	Zo	8388
18	*Lepismium lumbricoides* (Lem.) Barthlott	HLC	Zo	8386
19	*Rhipsalis baccifera* (J.S. Muell.) Stearn	HLC	Zo	8387
20	*Rhipsalis cereuscula* Haw.	HLC	Zo	8383
21	*Rhipsalis teres* (Vell.) Steud.	HLC	Zo	8384
22	*Rhipsalis trigona* Pfeiff.	HLC	Zo	8381
COMMELINACEAE				
23	*Tradescantia albiflora* Kunth	HLA	Zo	8391
ORCHIDACEAE				
24	*Acianthera recurva* (Lindl.) Pridgeon & M.W. Chase	HLC	An	8402
25	*Acianthera nemorosa* (Barb. Rodr.) F. Barros	HLC	An	8403
26	*Acianthera saundersiana* (Rchb.f.) Pridgeon & M.W.Chase	HLC	An	8401
27	*Anathallis obovata* (Lindl.) Pridgeon & M.W. Chase	HLC	An	8452
28	*Baptistonia lietzei* (Regel) Chiron & V.P.Castro	HLC	An	8407
29	*Bulbophyllum epiphytum* Barb. Rodr.	HLC	An	8374
30	*Bulbophyllum plumosum* (Barb.Rodr.) Cogn.	HLC	An	8409
31	*Bulbophyllum chloroglossum* Rchb.f. & Warm.	HLC	An	8408
32	*Epidendrum rigidum* Jacq.	HLC	An	8453
33	*Octomeria crassifolia* Lindl.	HLC	An	8472
34	*Octomeria palmyrabellae* Barb. Rodr.	HLC	An	8442
35	*Octomeria gracilis* Lodd. ex Lindl.	HLC	An	8440
36	*Oeceoclades maculata* (Lindl.) Lindl.	HLA	An	8400
37	*Ornithocephalus myrticola* Lindl.	HLC	An	8443
38	*Polystachya estrellensis* Rchb. f.	HLC	An	8467
39	*Polystachya foliosa* (Lindl.) Rchb.f.	HLC	An	8471
40	*Rodriguezia decora* Rchb. f.	HLC	An	8441
41	*Rodriguezia* sp.	HLC	An	8413
PIPERACEAE				
42	*Peperomia glabella* (Sw.) A. Dietr.	HMP	Zo	8475
43	*Peperomia rotundifolia* (L.) Kunth	HLC	Zo	8449
44	*Peperomia tetraphylla* (G. Forst.) Hook. & Arn.	HLC	Zo	8447
45	*Peperomia trineura* Miq.	HLC	Zo	8446
POLYPODIACEAE				
46	*Campyloneurum* sp.	HLC	An	8517
47	*Microgramma tecta* (Kaulf.) Alston	HLC	An	8524
48	*Microgramma persicariifolia* (Schrad.) C. Presl	HLC	An	8520
49	*Microgramma squamulosa* (Kaulf.) de la Sota	HLC	An	8534

Table 2 - Contents...

Family	Species	EC	Disp.	Reg.
POLYPODIACEAE				
50	*Peclima filicila* (Kaulf.) M.G. Price	HLC	An	8527
51	*Pleopeltis astrolepis* (Liebm.) E. Fourn.	HLC	An	8519
52	*Pleopeltis hirsitissima* (Raddi) de la Sota	HLC	An	8521
53	*Pleopeltis pleopeltifolia* (Raddi) Alston	HLC	An	8525
54	*Pleopeltis sqialida* (Vell.) de la Sota	HLC	An	8436
55	*Serpocailon latipes* (Langsd. & L. Fisch.) A.R. Sm.	HLF	An	8423
PTERIDACEAE				
56	*Vittaria lineata* (L.) Sm.	HLC	An	8454

The richness of epiphytic species found in the area can be considered close to that observed in the surveys in Semideciduous Seasonal Forest carried out by Rogalski and Zanin (2003) who found 70 species, by Giongo and Waechter (2004) who sampled 57 species and by Cervi and Borgo (2007) who found 56 species. This can be considered higher than that observed in the surveys carried out by Aguiar et al. (1981), who sampled 17 species, by Dislich and Mantovani (1998), with 34 species, by Borgo et al. (2002), with 32 species, by Breier (2005), with 25 species and by Dettke et al. (2008), with 29 species, still in areas of Semideciduous Seasonal Forest.

When compared to floristic surveys carried out in Cerrado areas, the vascular epiphytic community in the downstream area of the Sorocaba/Médio Tietê basin can be considered rich, especially when looking at the data from Breier (2005), who sampled 16 species, Ishara et al. (2008), seven species, Joanitti et al. (2010), 16 species and Bataghin et al. (2012b), who found 29 species. The presence of epiphytic species in ecotone areas between two forest formations is generally greater than both adjacent communities (BONNET et al., 2011). However, Ulhmann (1997), studying tree communities, postulated the existence of a "notorious invasion" of the forest over the cerrado. This competition, in the ecotone environment between the seasonal forest and the cerrado, can generate a community with a new composition, with species shared by both phytophysiognomies, but not more diverse than the phytophysiognomy with the greatest diversity. Many epiphytic species may be excluded, not only because of existing interspecific competition, but also because of abiotic factors (greater luminosity and less water supply) which affect the ecotone differently from the adjacent plant formations. Another relevant fact is that the Cerrado formations have a smaller number of epiphytic species than the seasonal forests and, for the most part, these species are characteristic of environments with a water deficit. Corroborating this trend of a low number of species, Bataghin et al. (2012b) observed 29 species in a Conservation Unit in a Cerrado area of more than 9,000 ha in the interior of Sao Paulo; in addition, Joanitti et al. (2010), in an area similar to the Cerrado, found 16 species and Breier (2005) and Ishara et al. (2008), sampled 16 species and seven species, respectively, in Cerrado areas.

It is important to note that the number of species sampled here (56 spp.) represents the total number of epiphytic species found in all the sites in the Downstream Area of the Sorocaba/Médio Tietê basin, which means that seven different forest fragments were surveyed (the detailed richness of each area will be presented in the course of this thesis). As such, the richness of the downstream area can be considered to be negligible, which is due to the dry climate characteristic of this part of the basin and also to all the anthropogenic actions to which this part of the basin is subjected, especially the massive reduction of forests and the removal of larger trees, which are fundamental for the maintenance of the vascular epiphytic community.

In addition, the specific richness of the area, as expected, can be considered low when compared to wetter forest formations. In dense ombrophilous forest, several authors report a higher number of epiphyte species, e.g. Kersten (2006) - 349 species; Breier (2005) - 161 species; Fontoura et al. (1997) - 293 species; Hertel (1950) - 101 species, Schütz-Gatti (2000) - 175 species and Petean (2003) - 97 species. This reinforces the idea of dependence on atmospheric humidity (Gentry and Dodson 1987a), since the acquisition and storage of water are the most important factors for epiphytic growth (ZOTS; HIETZ, 2001).

In the survey of the Ecotone between Semideciduous Seasonal Forest and Cerrado in the downstream area of the Sorocaba/Médio Tietê river basin, the epiphytic families with the highest species richness were: Orchidaceae (18 species), Bromeliaceae (12 species), Polypodiaceae (10 species) and Cactaceae (eight species). The Commelinaceae and Pteridaceae families had only one species. The distribution of epiphytic species in the ecological categories (Figure 3), according to the relationship with the forophyte proposed by Benzing (1990), showed a predominance of characteristic holoepiphytes with 48 species (86%), followed by facultative holoepiphytes (5%), accidental holoepiphytes (4%) and primary hemiepiphytes with two species (4%) and secondary hemiepiphytes with only one species. The predominance of characteristic holoepiphytes has been observed in Semideciduous Seasonal Forest in several studies (PINTO et al., 1995; DISLICH; MANTOVANI, 1998; ROGALSKI; ZANIN, 2003; CERVI; BORGO, 2007; DETTKE et al., 2008; BATAGHIN et al, 2010b), in Cerrado areas (BREIER, 2005; BATAGHIN et al., 2012b) and in other forest formations, such as Mixed Ombrophilous Forest (DITTRICH et al., 1999) and in restinga areas (WAECHTER, 1992; KERSTEN; SILVA, 2001).

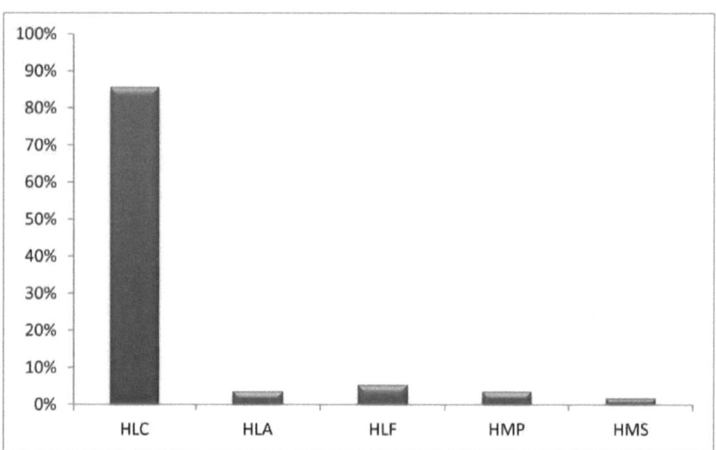

Figura 3: Distribution of the epiphytic vascular species of the Seasonal Semideciduous Forest in the downstream area of the Sorocaba/Médio Tietê hydrographic basin in the ecological categories proposed by Benzing (1990) - HLC: characteristic holoepiphytes; HLF: facultative holoepiphytes; HLA: accidental holoepiphytes; HMP: primary hemiepiphytes; HMS: secondary hemiepiphytes.

The classification of the vascular epiphytes of the Downstream Area of the hydrographic basin, in terms of dispersal syndrome, highlighted 64.3% of anemochoric species and 35.7% zoochoric. Dispersal strategy is an important factor in the success of epiphytic synopsis (GENTRY; DODSON, 1987a), and notably anemochory has predominated as a dispersal syndrome among epiphytic species (BENZING, 1987; BREIER, 2005; DETTKE et al., 2008; MENINI-NETO et al., 2009; GERALDINO et al., 2010). This high percentage of anemochory is a reflection of the large number of orchids, ferns and bromeliads (in the latter case, especially the genus *Tillandsia)* recorded in the study area.

The Orchidaceae, Bromeliaceae and Polypodiaceae families are responsible for 71% (40 spp.) of the species found in the floristic survey, a percentage very similar to that found by Kersten (2006); in addition, these families are considered to be the richest in epiphytes worldwide (MADISON, 1977; KRESS 1986; GENTRY; DODSON, 1987b; BENZING, 1990). The Cactaceae family also deserves to be highlighted in the research area, because although it accounts for around 0.5% of the world's epiphytic species (MADISON, 1977; BENZING, 1990) and 3% of Brazilian epiphytes (Kersten 2006), in the study area it had eight species, making up 14%. The resistance of the Cactaceae to periods of water stress due to their "relatively accentuated xeromorphism" (COUTINHO, 1962), an adaptation which probably offers benefits in relation to periods of water deficit, may be responsible for the representativeness of this family in the area in question.

An important fact concerns the Orchidaceae family, which is the richest worldwide (MADISON, 1977; KRESS, 1986; BENZING, 1990), in the Neotropics (GENTRY; DODSON, 1987a) and in Brazil (KERSTEN, 2006). This family had 18 species, a smaller number than those found by Rogalski and Zanin (2003), who found 38 species in the Semideciduous Seasonal Forest (FES), but very similar to the results obtained by Giongo and Waechter (2004) - 16 species (FES), by Cervi and Borgo (2007) - nine species (FES) and Bonnet et al. (2011) - 16 species (FES), and a greater number than the works published by Dislich and Mantovani (1998) - 6 species (FES), Breier (2005) - 3 species (FES), Dettke et al. (2008) - 3 species (FES); Bataghin et al. (2010) - 2 species (FES), Breier (2005) - 3 species (Cerrado) and Bataghin et al. (2012b) - 8 species (Cerrado). Stancato et al. (2002) suggest that high light intensity can reduce the growth and development of orchids, but it is not clear that the climatic variations characteristic of the study area have any influence on these numbers. Ditt (2002) highlights anthropogenic interference, microclimate changes resulting from environmental modification and the possible illegal collection of specimens (especially those of commercial - ornamental - interest) as factors that can lead to a reduction in the number of individuals, contributing to a decrease in biological diversity and environmental degradation. However, the difficult access to the study areas (most of which are steep) may have contributed to the reduction in possible illegal collection of species, and may be related to the higher number of species for the Orchidaceae family in this part of the river basin.

Distribution of vascular epiphytes in the sampling sites in the Downstream Area

Analysis of the vascular epiphytes of the Core Site of the Semideciduous Seasonal Forest/Cerrado Ecotone in the Downstream Area of the Sorocaba/Médio Tietê Basin

The epiphyte survey carried out at the Core Site of the Downstream Area of the Sorocaba/Médio Tietê Hydrographic Basin took place at the Barreiro Rico Ecological Station, an integral protection conservation unit, located between UTM coordinates 793.598 and 7.489.761 in zone 22 South, which covers 292.82 ha in the municipality of Anhembi- SP. The area was established as a UC at the end of 2006 by State Decree 51.381, with the aim of conserving the enclave and the surrounding forest (SÂO PAULO, 2006). The region's climate is of the Cwa type, according to the Koppen classification system, tropical with the rainy season from September to March and the dry season from April to August. The altitude of the area varies between 500 and 580 m. Locally, the remnant studied can be considered large. The region of the Barreiro Rico Ecological Station can be considered an area of extreme biological importance and a priority for conservation (CONSERVATION INTERNATIONAL DO

BRASIL, 2000).

The floristic survey of Sitio Core recorded 25 species, belonging to 15 genera and six families (Table 3). The Shannon index of the site was H' = 2.773, the equability (J) was 0.861 and the Margalef richness (d) was 3.406.

Table 3 - List of vascular epiphyte species found in the Core Site of the Downstream Area (Barreiro Rico Ecological Station) of the Sorocaba/Médio Tietê hydrographic basin and their respective Ecological Categories (EC) - HLC: Holoepiphyte characteristic; HLF: Holoepiphyte facultative; HLA: Holoepiphyte accidental; HMP: Hemiepiphyte primary: HMS: Hemiepiphyte secondary. Dispersal form (Disp.) - Zo: Zoochoric; Na: Anemochoric. Reg: HUFSCar herbarium registration number (Im: Digital image).

Family	Species	EC	Disp.	Reg.
ARACEAE				
1	Philodendron appendiculatum Nadruz & S.J. Mayo	HMS	Zo	8487
2	Philodendron bipinnatifidum Schott	HMP	Zo	Im
BROMELIACEAE				
3	Aechmea distichantha Lem.	HLF	Zo	8503
4	Tillandsia recurvata (L.) L.	HLC	An	8429

Keep going...

Table 3 - Continuation...

Family	Species	EC	Disp.	Reg.
BROMELIACEAE				
5	*Tillandsia stricta* Sol. ex Sims	HLC	An	8428
6	*Tillandsia tricholepis* Baker	HLC	An	8430
7	*Vriesea bituminosa* Wawra	HLC	An	8515
CACTACEAE				
8	*Epiphyllum phyllanthus* (L.) Haw.	HLC	Zo	8474
9	*Lepismium cruciforme* (Vell.) Miq.	HLC	Zo	8388
10	*Lepismium lumbricoides* (Lem.) Barthlott	HLC	Zo	8386
11	*Rhipsalis baccifera* (J.S. Muell.) Stearn	HLC	Zo	8387
12	*Rhipsalis cereuscula* Haw.	HLC	Zo	8383
ORCHIDACEAE				
13	*Baptistonia lietzei* (Regel) Chiron & V.P.Castro	HLC	An	8407
14	*Bulbophyllum plumosum* (Barb.Rodr.) Cogn.	HLC	An	8409
15	*Bulbophyllum epiphytum* Barb. Rodr.	HLC	An	8374

16	*Octomeria crassifolia* Lindl.	HLC	An	8472
17	*Ornithocephalus myrticola* Lindl.	HLC	An	8443
18	*Rodriguezia* sp.	HLC	An	8413
PIPERACEAE				
19	*Peperomia rotundifolia* (L.) Kunth	HLC	Zo	8449
20	*Peperomia tetraphylla* (G. Forst.) Hook. & Arn.	HLC	Zo	8447
POLYPODIACEAE				
21	*Microgramma persicariifolia* (Schrad.) C. Presl	HLC	An	8520
22	*Microgramma squamulosa* (Kaulf.) de la Sota	HLC	An	8534
23	*Microgramma tecta* (Kaulf.) Alston	HLC	An	8524
24	*Pleopeltis hirsutissima* (Raddi) de la Sota	HLC	An	8521
25	*Pleopeltis pleopeltifolia* (Raddi) Alston	HLC	An	8525

The richness of epiphytic species found in the Barreiro rico Ecological Station (Sitio Core - Area Jusante) can be considered low, especially if we remember the studies carried out in Semideciduous Seasonal Forest by Rogalski and Zanin (2003) who found 70 species, by Giongo and Waechter (2004) who sampled 57 species and by Cervi and Borgo (2007) who found 56 species, although it can be considered similar to that observed by Aguiar et al. (1981), who sampled 17 species, by Dislich and Mantovani (1998), with 34 species, by Borgo et al. (2002), with 32 species, by Breier (2005), with 25 species and by Dettke et al. (2008), with 29 species.

When compared to studies carried out in Cerrado areas, the richness of Sitio Core da Area Jusante can be considered similar when compared to the results of Bataghin et al. (2012b), who found 29 species, and superior when considering the data of Breier (2005), who sampled 16 species, Ishara et al. (2008), seven species, and Joanitti et al. (2010), 16 species.

The vascular epiphyte richness of Sitio Core may have been influenced by interference during the period when the area had not been declared a Conservation Unit, a fact that only occurred in 2006 (SÂO PAULO, 2006), since the time that has elapsed since then can be considered short in ecological terms for the recovery of the vascular epiphyte community. Several authors have reported the loss of epiphytic diversity as a result of human interference in environments (BARTHLOTT et al., 2001; WOLF, 2005; BATAGHIN et al., 2008; DETTKE et al., 2008); in addition, Engwald et al. (2000) reported that the lack of structural complexity in anthropized forests means that the epiphytic community, in addition to being little diverse, takes much longer to (re)establish. It should be noted that the Conservation Unit shows signs of anthropic activities still taking place in its interior, especially hunting activities.

At Sitio Core, Orchidaceae was the richest family with six species. Bromeliaceae, Cactaceae and Polypodiaceae had five species each and the Araceae and Piperaceae families had only two species each. Characteristic holoepiphytes were dominant with 88% of the species, followed by primary hemiepiphytes, secondary hemiepiphytes and facultative holoepiphytes, all with 4% of the species. Accidental holoepiphytes were not recorded in the core site of the Downstream Area. As for the dispersal syndrome, 10 species were zoochoric and 15 anemochoric. For both ecological categories and dispersal syndromes, these are the patterns usually observed in dry forests.

The quantitative analysis recorded 25 species and highlighted two Polypodiaceae as the most important species at Sitio Core (Table 4) - *Microgramma squamulosa* (Figure 4) and *Microgramma tecta* were responsible for more than 30% of the epiphytic importance value (VIE) recorded at this site. The former had a VIE of 19.6 and an average score of 1.72, being recorded in almost 50% of the forophytes and almost 25% of the strata. The second species had a VIE of 10.63 and an average score of 1.67, as well as occurring in 27.8% of the forophytes and 13.52% of the strata. This contributed to the Polypodiaceae family, with 40.4% of the VIE, being highlighted as the most important in Sitio Core.

Table 4 - Vascular epiphytes of the Core Site of the Downstream Area in the Sorocaba Médio Tietê hydrographic basin, classified according to the value of epiphytic importance - nr: absolute number of occurrences in the strata; far: absolute frequency in the strata; ni: absolute number of occurrences in the forophytic individuals; fai: absolute frequency in the forophytic individuals; vt (total value): sum of the abundance estimates; vie: value of epiphytic importance; note: average note obtained.

Species	nr	far	ni	fai	vt	vie	note
Microgramma squamulosa	131	24.3	43	47.8	225	19.60	1.72
Microgramma tecta	73	13.5	25	27.8	122	10.63	1.67
Lepismium lumbricoides	55	10.2	24	26.7	119	10.37	2.16
Peperomia rotundifolia	32	5.9	13	14.4	64	5.57	2.00
Rhipsalis cereuscula	25	4.6	10	11.1	59	5.14	2.36
Epiphyllum phyllanthus	30	5.6	13	14.4	58	5.05	1.93
Tillandsia tricholepis	32	5.9	11	12.2	57	4.97	1.78
Ornithocephalus myrticola	31	5.7	14	15.6	57	4.97	1.84
Tillandsia recurvata	31	5.7	11	12.2	56	4.88	1.81
Pleopeltis pleopeltifolia	37	6.9	12	13.3	55	4.79	1.49
Philodendron bipinnatifidum	22	4.1	10	11.1	45	3.92	2.05
Lepismium cruciforme	23	4.3	11	12.2	40	3.48	1.74
Vriesea bituminosa	21	3.9	11	12.2	38	3.31	1.81
Pleopeltis hirsutissima	24	4.4	10	11.1	34	2.96	1.42
Microgramma persicariifolia	17	3.1	7	7.8	28	2.44	1.65

Philodendron appendiculatum	11	2.0	6	6.7	21	1.83	1.91
Bulbophyllum plumosum	7	1.3	4	4.4	13	1.13	1.86
Baptistonia lietzei	6	1.1	4	4.4	11	0.96	1.83
Octomeria crassifolia	4	0.7	3	3.3	10	0.87	2.50
Rodriguezia sp.	6	1.1	3	3.3	9	0.78	1.50
Bulbophyllum epiphytum	3	0.6	3	3.3	7	0.61	2.33
Rhipsalis baccifera	4	0.7	2	2.2	5	0.44	1.25
Aechmea distichantha	2	0.4	1	1.1	5	0.44	2.50
Tillandsia stricta	3	0.6	1	1.1	5	0.44	1.67
Peperomia tetraphylla	2	0.4	1	1.1	5	0.44	2.50

The species of the Cactaceae family, especially *Lepismium Iumbricoides* with a VIE = 10.37 and an average score of 2.16 and *Rhipsalis cereuscula* with a VIE = 5.14 and an average score of 2.36, helped make this family the second most important for this site, with a total VIE of 24.48. The Bromeliaceae family had the third highest importance value - 14.02 - and the Orchidaceae family, although it had the highest number of species, had a low epiphytic importance value, with a VIE of 4.36. The low importance value of the Orchidaceae family reflects the fragility of species and individuals of this family in terms of surviving climatic variations and possible anthropic pressures in the area.

Figura 4: Microgramma squamulosa (Kaulf.) de la Sota (Polypodiaceae), the most important epiphytic species in the Downstream Core Site.

The representativeness, in terms of number of species, of the Orchidaceae and Polypodiaceae families may be related to the fact that they are considered the richest in epiphytes both in the world (GENTRY; DODSON, 1987b) and in Brazil (KERSTEN, 2006). As for the Bromeliaceae and Cactaceae families, Dislich and Mantovani (1998) highlight their neotropical endemism, a fact that may be responsible for the numbers presented here. In

39

addition, according to Scheinvar (1985), the genus *Rhipsalis* (Cactaceae), for example, has its dispersal center in the south and southeast of Brazil, which favors the occurrence of species from this family in the area studied. However, the greater abundance of species with resistance to periods of water deficit is noteworthy, as is the case with Polypodiaceae *(Microgramma)* and Cactaceae (*Lepismium*) which, in the first case, are well adapted to environments with more sunlight and, in the second case, because they are succulents, are highly resistant to dry periods.

The richness of epiphytic species, although small, cannot be considered low for the study area in question, especially if we look at the richness data for cerrado areas (JOANITTI et al., 2010; BATAGHIN et al., 2012b). Although the occurrence of few species of epiphytes is a characteristic of forests that suffer from anthropic interference (BATAGHIN et al., 2008; BATAGHIN et al., 2010), the figures obtained in this study do not allow us to clearly state that the possible anthropic actions that affected the Barreiro Rico UC before its establishment as a conservation unit, which only occurred in 2006 (and possibly still do), have an influence on the vascular epiphytic community of this site.

However, the concentration of species in the Bromeliaceae, Cactaceae and Polypodiaceae families, common in environments with greater luminosity and lower humidity (DETTKE et al., 2008), may be related to environmental factors characteristic of Sitio Core, such as marked climatic variations, including prolonged periods of water deficit, since some species of these families have adaptations that enable them to survive in these conditions. In addition, the Orchidaceae family has the largest number of species among the richest families in the area, a fact that does not allow us to clearly attest to possible occasional anthropogenic interference in the area, especially in the collection of species of ornamental/economic interest, such as orchids, although it should be noted that the orchid species found in the area are generally small and have little ornamental appeal (Figure 5). This indicates a moderate anthropogenic influence and a greater importance of abiotic factors on the epiphytic community in the core site of the downstream area.

Figura 5: Ornithocephalus myrticola Lindl. (Orchidaceae), a small species with less ornamental appeal.

The distribution of epiphytes in the forophyte strata at Sitio Core (Figure 6) showed the base of the canopy as the stratum with the highest epiphyte abundance, with an abundance value (VA) of 335. The second most abundant stratum was the inner canopy with a VA of 249, followed by the upper canopy with VA = 240, the middle canopy with VA = 167, the lower canopy with VA = 103 and the outer canopy with VA = 54.

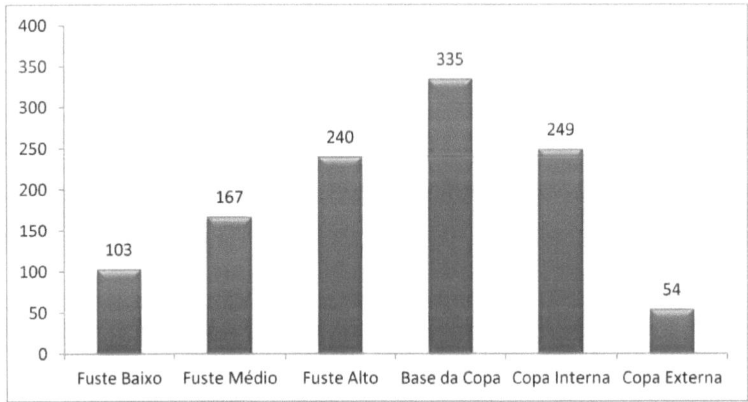

Figura 6: Distribution of the abundance of epiphytic vascular species among the forophytic strata in the Core Site of the Downstream Area in the Sorocaba/Médio Tietê hydrographic basin.

Statistical analysis applied to the vertical distribution of vascular epiphytes in the forophytic strata, based on species abundance, revealed the low stem and the outer canopy as the most different strata. The low stem differed from three strata (Table 5), while the outer canopy was significantly different in four comparisons.

The vertical distribution pattern observed at Sitio Core may be related to what is known as "vertical evolution", in which epiphytes move to lower spaces in search of more light and conditions for acquiring water and nutrients (KIRA; YODA, 1989; BENZING, 1990).

Table 5: Statistical analysis of the richness and abundance of vascular epiphytes among the forophytic strata in the Core Site of the Downstream Area in the Sorocaba/Médio Tietê watershed.

^^^^^^Wealth Abundance^^^^	Beam Bass	Beam Medium	High stem	World Cup base	Cup Internal	Cup External
Low stem		0,264	**0,02025**	**0,008096**	0,5828	**0,03969**
Medium stem	0,126		0,2169	0,1134	0,5754	**0,001878**
High stem	**0,031**	0,167		0,7265	0,07366	**2,80E-05**
World Cup base	**0,008**	**0,042**	0,186		**0,03348**	7,74E-06
Internal Cup	**0,046**	0,177	0,464	0,229		**0,009757**
External Cup	0,149	**0,017**	**0,005**	**0,002**	**0,011**	

However, the lower number of individuals in the outer canopy is a sign of the limitation of epiphytes by more intense abiotic factors, such as higher luminosity and lower moisture availability, and the lack of light in the lower part of the forest tends to reduce the number of individuals. Also noteworthy is the importance of places that can serve as deposits of suspended soil, especially the insertion points of branches at the base of the canopy and in the inner canopy, which serve as a supplement of water and nutrients for vascular epiphytes (KROMER et al., 2007). Although there is a variation in the composition of the species that establish themselves in the low stem, they are distributed in all strata of the forophyte, a possible indication of the limitation of more demanding species even in the lower part of the forest, where there are better conditions for acquiring water, in addition to the establishment of few species adapted to conditions of greater luminosity in the outer canopy.

Analysis of the vascular epiphytes at Sitio Rèplica I in the downstream area of the Sorocaba/Médio Tietê basin

The survey of vascular epiphytes at Sitio Rèplica I in the Ecotone of the Semideciduous Seasonal Forest and the Cerrado in the Downstream Area of the watershed was carried out on a private rural property in the municipality of Sao Manuel-SP. The forest remnant of approximately 95 ha, located at UTM coordinates 754.994 and 7.485.482 in zone 22 South, is characterized by being an ecotone area between the Semideciduous Seasonal Forest and the Cerrado, as well as one of the largest forest fragments in the municipality. The steep slope of the area (Figure 7) in the forest remnant played a fundamental role in maintaining the vegetation cover. Although the relief of the area is very steep, there were strong indications of anthropic presence, such as the existence of trails, waiting areas commonly used for hunting, as well as hunting traps known as "canhao" (a type of firearm designed to shoot animals that trip over a line attached to the trigger). The surrounding areas are characterized by agricultural use, with the cultivation of sugar cane, coffee and eucalyptus.

Figura 7: Photograph of Sitio Rèplica I in the Downstream Area, showing the slope of the forest fragment.

The floristic analysis of the area revealed the presence of 11 species, belonging to eight genera and five families (Table 6). The Shannon index for the area was H' = 2.091, the equability (J) was 0.871 and the Margalef richness (d) was 1.381.

Table 6 - List of vascular epiphyte species found at Sitio Rèplica I in the downstream area (Fazenda Sao Manuel) of the Sorocaba Mèdio Tietê river basin and their respective Ecological Categories (EC) - HLC: Characteristic holoepiphyte; HLF: Facultative holoepiphyte; HLA: Accidental holoepiphyte; HMP: Primary hemiepiphyte; HMS: Secondary hemiepiphyte. Dispersal form (Disp.) - Zo: Zoochoric; An: Anemochoric. Reg: HUFSCar herbarium record number (Im: Digital image).

N	Family	Species	EC	Disp.	Reg.
1	Araceae	*Philodendron bipinnatifidum* Schott ex Endl.	HMP	Zo	Im
2	Bromeliaceae	*Tillandsia recurvata* (L.) L.	HLC	An	8429
3	Bromeliaceae	*Tillandsia tricholepis* Baker	HLC	An	8430
4	Cactaceae	*Epiphyllum phyllanthus* (L.) Haw.	HLC	Zo	8474
5	Cactaceae	*Lepismium cruciforme* (Vell.) Miq.	HLC	Zo	8388
6	Cactaceae	*Lepismium lumbricoides* (Lem.) Barthlott	HLC	Zo	8386

7	Cactaceae	*Rhipsalis teres* (Vell.) Steud.	HLC	Zo	8384
8	Piperaceae	*Peperomia rotundifolia* (L.) Kunth	HLC	Zo	8449
9	Polypodiaceae	Microgramma persicariifolia (Schrad.) C. Presl	HLC	An	8520
10	Polypodiaceae	Pleopeltis pleopeltifolia (Raddi) Alston	HLC	An	8525
11	Polypodiaceae	*Pleopeltis squalida* (Vell.) de la Sota	HLC	An	8436

Although there are no surveys of the diversity of vascular epiphytes in ecotone regions similar to the study area, the richness of the area can be considered low when compared to studies carried out in Semideciduous Seasonal Forest, such as. e.g. Rogalski and Zanin (2003) - 70 species; Giongo and Waechter (2004) - 57 species; Cervi and Borgo (2007) - 56 species; Dislich and Mantovani (1998) - 34 species; Borgo et al. (2002) - 32 species; Breier (2005) - 25 species; Dettke et al. (2008) - 29 species; Bataghin et al. (2010) - 21 species and Bonnet et al. (2011) - 60 species. When compared to Cerrado areas, the richness can be considered lower than that observed by Bataghin et al. (2012b) - 29 species and similar to the studies by Breier (2005) - 16 species, Ishara et al. (2008) - seven species and Joanitti et al. (2010) - 16 species.

The Règlica I site showed lower richness than the other sites (Core and Replicas II and III) sampled in the downstream area of the Sorocaba/Médio Tietê basin. The low number of epiphytic species may be related to the lack of complexity of the forest, either due to the absence of a favorable microclimate (constant strong wind due to the steep slope of the area) or even due to the reduction in the number of forophytes, especially larger trees, since these factors are responsible for reducing epiphytic diversity (ENGWALD et al., 2000; BARTHLOTT et al., 2001; DETTKE et al., 2008).

The Cactaceae family, cited by Dettke et al. (2008) as one of the richest in an altered fragment of Semideciduous Seasonal Forest in the city of Maringà (PR), was responsible for 36% of the epiphytic species in the area (four species), followed by the Polypodiaceae family with three species (27%), which is also cited by the above authors. The Bromeliaceae family had two species. Characteristic holoepiphytes predominated at this site, accounting for almost 91% of the species, followed by primary hemiepiphytes with 9.1% of the species; no facultative holoepiphytes, accidental holoepiphytes or secondary hemiepiphytes were recorded. This predominance of characteristic holoepiphytes, although common to various forest formations, is accentuated in dry forests or forests with a greater degree of anthropogenic influence. This is related to the presence of adaptations that give characteristic holoepiphyte species greater resistance to periods of water deficit.

Another important result for this site is the classification of species by dispersal syndrome.

Six species were found to be zoochorically dispersed and only five anemochorically dispersed. This result was not expected, because not only do most of the epiphytes (about 2/3 of them) show anemochory as a dispersal syndrome, but the vegetation (forest with low structural complexity) and climatic characteristics of this small forest fragment, in theory, would not favor zoochoric dispersal, which was recorded here in more than 50% of the species. However, the isolation of the forest fragment by the agricultural matrix may represent a barrier to the dispersal of anemochoric species, and the consequent arrival of new species with this syndrome in the fragment, justifying the reduction in the expected proportion.

In the quantitative analysis, 11 species were recorded, of which two Bromeliaceae were the most prominent species at Sitio Réplica I (Table 7).

Table 7 - Vascular epiphytes from Sitio Réplica I in the Downstream Area of the Sorocaba Médio Tietê river basin, classified according to epiphytic importance value - nr: absolute number of occurrences in the strata; far: absolute frequency in the strata; ni: absolute number of occurrences in the forophytic individuals; fai: absolute frequency in the forophytic individuals; vt (total value): sum of the abundance estimates; vie: epiphytic importance value; nota: average score obtained.

Species	nr	far	ni	fai	vt	vie	note
Tillandsia recurvata	176	32.59	58	64.44	314	22.48	1.78
Tillandsia tricholepis	155	28.70	48	53.33	284	20.33	1.83
Rhipsalis teres	94	17.41	41	45.56	200	14.32	2.13
Peperomia rotundifolia	76	14.07	40	44.44	187	13.39	2.46
Pleopeltis squalida	71	13.15	25	27.78	132	9.45	1.86
Lepismium cruciforme	29	5.37	13	14.44	70	5.01	2.41
Microgramma persicariifolia	30	5.56	12	13.33	62	4.44	2.07
Epiphyllum phyllanthus	25	4.63	14	15.56	61	4.37	2.44
Pleopeltis pleopeltifolia	25	4.63	12	13.33	39	2.79	1.56
Philodendron bipinnatifidum	15	2.78	7	7.78	35	2.51	2.33
Lepismium lumbricoides	6	1.11	3	3.33	13	0.93	2.17

Tillandsia recurvata had an epiphytic importance value (EVI) of 22.48 and an average score of 1.78, occurring in almost 65% of the forophytes and 33.6% of the strata, making it the most important species at this site (Figure 8). *Tillandsia tricholepis* had a VIE of 20.33, an average score of 1.83 and was recorded in 53% of the forophytes and 28.7% of the strata. *Rhipsalis teres*, recorded in 44.4% of the forophytes and 17.4% of the strata, with an average score of 2.13, had a VIE of 14.32 and was the third most important species. *Peperomia rotundifolia* with a VIE of 13.39 and an average score of 2.46 was the fourth most important species at

Sitio Réplica I. These four species were responsible for more than 70% of the VIE of this site.

Figura 8: Tillandsia recurvata (L.) L. (Bromeliaceae), the most abundant species in Sitio Règlica I in the Downstream Area, typical of areas (strata) with higher luminosity and low water availability.

The Bromeliaceae family, although it only had two species in the area, was also responsible for almost 43% of the epiphytic importance value. The Cactaceae family had four species and was the second most important on the site with a VIE of 24.62. The concentration of abundance in a few species is typical of impacted areas (BATAGHIN et al., 2010) and may be related to models of pre-emptying of niches (MAY, 1975). At this site, there is a tendency for species to be concentrated in a few families, which are predominant in areas under anthropogenic pressure or, possibly, in areas with a large water deficit.

The distribution of epiphytes in the forophyte strata (Figure 9) showed that the base of the canopy, with an abundance value (VA) of 393, was the stratum with the highest epiphyte abundance. The second most abundant stratum was the inner canopy with VA = 321, followed by the high canopy with VA equal to 225, the outer canopy with VA = 199, the middle canopy with VA = 137 and the low canopy with VA = 122.

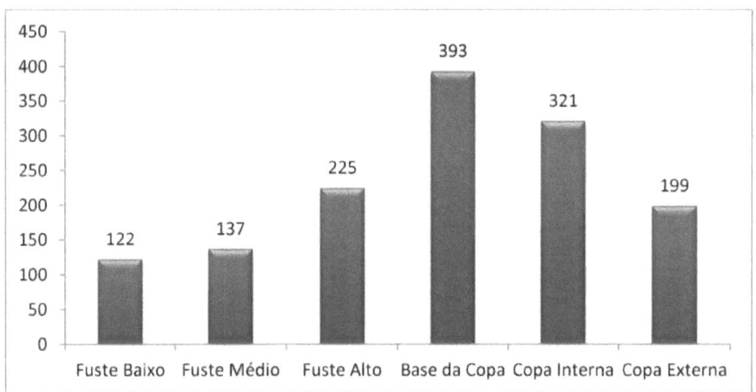

Figura 9:   Distribution of the abundance of epiphytic vascular species among the forophytic strata at Sitio Réplica I in the Downstream Area of the Sorocaba/Médio Tietê hydrographic basin.

The distribution of vascular epiphytes in the forophytic strata at Sitio Réplica I (Table 8), based on the abundance of species, showed that only the base of the canopy was significantly different from the low stem and the mid stem. The vertical distribution of richness was significantly different between the low stem and the three strata, while the outer canopy showed a different floristic composition from the high stem and the base of the canopy.

Table 8: Statistical analysis of the richness and abundance of vascular epiphytes among the forophytic strata at Sitio Réplica I in the Downstream Area of the Sorocaba/Médio Tietê hydrographic basin.

^^^^^^Wealth Abundance^^^^	Beam Bass	Beam Medium	Beam High	Cup base	Cup Internal	Cup External
Low stem		0,2178	**0,0034**	**0,0032**	**0,0134**	0,4084
Medium stem	0,453		0,0675	0,0675	0,1933	0,7049
High stem	0,189	0,188		0,9475	0,5811	**0,0281**
Cup base	**0,035**	**0,03**	0,082		0,5811	**0,0281**
Internal Cup	0,104	0,102	0,233	0,323		0,0905
External Cup	0,316	0,339	0,425	0,12	0,24	

The tendency to concentrate the abundance of epiphytes at Sitio Réplica I in the intermediate strata, especially at the base of the canopy, where there is possibly an intermediate climatic situation between water stress and access to light, in addition to the presence of branch insertion points which may favor the accumulation of suspended soil, this has been the same behavior observed in the epiphyte community at the core site in the downstream area of the Sorocaba/Médio Tietê basin and also in other studies in moist forests (KROMER et al., 2007; ROGALSKI; ZANIN, 2003) and also in dry forests (BREIER, 2005). The differences in richness are related to the occurrence of exclusive species in certain strata of the forophyte, where common species, widely distributed over their host trees, predominate, especially in the low stem and outer canopy. However, the absence of a significant difference in terms of richness between the low stem and the outer canopy corroborates the idea that the environmental conditions within this site allow the same species to grow in these strata, which leads us to believe that the differences in relation to the intermediate strata may be related to the presence of a larger area for installation, more favorable microclimatic conditions or even interspecific competition within this part of the forophyte.

Analysis of the vascular epiphytes at Sitio Règlica II in the downstream area of the Sorocaba/Médio Tietê river basin

Sitio Réplica II in the downstream area of the Sorocaba/Médio Tietê watershed was established in a forest area on a private property in the municipality of Botucatu, São Paulo. The forest fragment, located at UTM coordinates 768.703 and 7.486.350 in zone 22 South, is a remnant of Cerrado on the banks of a small stream, where Cerrado, paludosa and gallery forest environments also appear. The area is approximately 460 ha in size and is the largest forest fragment of this phytophysiognomy in the municipality and in the Sorocaba/Médio Tietê watershed area. The area's surroundings are characterized by orange cultivation, which relies on the constant application of pesticides, which influences the fragment's vascular epiphytic community, mainly by eliminating pollinators.

At Sitio Réplica II in the downstream area of the Sorocaba/Médio Tietê basin, 24 species were found, belonging to 16 genera and seven families (Table 9). The site's Shannon index was $H' = 1.924$, equability (J) was 0.679 and Margalef's richness (d) was 2.266.

Table 9 - List of vascular epiphyte species found at Sitio Règlica II in the downstream area (Fazenda Sao Luiz) of the Sorocaba Mèdio Tietê hydrographic basin and their respective Ecological Categories (EC) - HLC: Holoepiphyte characteristic; HLF: Holoepiphyte facultative; HLA: Holoepiphyte accidental; HMP: Hemiepiphyte primary; HMS: Hemiepiphyte secondary. Dispersal form (Disp.) - Zo: Zoochoric; An: Anemochoric. HUFSCar herbarium record number (Im: Digital image).

N	Family	Species	EC	Disp.	Reg.
1	Araceae	*Philodendron bipinnatifidum* Schott	HMP	Zo	Im
2	Bromeliaceae	*Acanthostachys strobilacea* (Schult. f.) Klotzsch	HLC	Zo	8431
3	Bromeliaceae	*Aechmea bromeliifolia* (Rudge) Baker	HLC	Zo	8432
4	Bromeliaceae	*Tillandsia funckiana* Baker	HLC	An	8458
5	Bromeliaceae	*Tillandsia recurvata* (L.) L.	HLC	An	8429

6	Bromeliaceae	*Tillandsia stricta* Sol. ex Sims	HLC	An	8428
7	Bromeliaceae	*Tillandsia tricholepis* Baker	HLC	An	8430
8	Cactaceae	*Epiphyllum phyllanthus* (L.) Haw.	HLC	Zo	8474
9	Cactaceae	*Lepismium lumbricoides* (Lem.) Barthlott	HLC	Zo	8386
10	Cactaceae	*Rhipsalis cereuscula* Haw.	HLC	Zo	8383
11	Orchidaceae	*Acianthera recurva* (Lindl.) Pridgeon & M.W. Chase	HLC	An	8402
12	Orchidaceae	*Acianthera nemorosa* (Barb. Rodr.) F. Barros	HLC	An	8403
13	Orchidaceae	*Anathallis obovata* (Lindl.) Pridgeon & M.W. Chase	HLC	An	8452
14	Orchidaceae	*Epidendrum rigidum* Jacq.	HLC	An	8453
15	Orchidaceae	*Oeceoclades maculata* (Lindl.) Lindl.	HLA	An	8400
16	Orchidaceae	*Polystachya foliosa* (Lindl.) Rchb.f.	HLC	An	8471
17	Piperaceae	*Peperomia glabella* (Sw.) A. Dietr.	HMP	Zo	8475
18	Polypodiaceae	*Microgramma tecta* (Kaulf.) Alston	HLC	An	8524
19	Polypodiaceae	*Microgramma squamulosa* (Kaulf.) de la Sota	HLC	An	8534
20	Polypodiaceae	*Pleopeltis hirsutissima* (Raddi) de la Sota	HLC	An	8521
21	Polypodiaceae	*Pleopeltis pleopeltifolia* (Raddi) Alston	HLC	An	8525
22	Polypodiaceae	*Pleopeltis squalida* (Vell.) de la Sota	HLC	An	8436
23	Polypodiaceae	*Serpocaulon latipes* (Langsd. & L. Fisch.) A.R. Sm.	HLF	An	8423
24	Pteridaceae	*Vittaria lineata* (L.) Sm.	HLC	An	8454

The richness of this site is higher than that found in studies carried out in the same type of forest by Breier (2005) - 16 species, Ishara et al. (2008) - seven species and Joanitti et al. (2010) - 16 species, and similar to the data obtained by Bataghin et al. (2012b) who found 29 epiphytic species in a Conservation Unit in a Cerrado area. This site showed similar richness to that observed in the core site of the downstream area (a Conservation Unit) and higher than that found in replicate site I, both sampled in this study. The presence and abundance of species that are more resistant to brighter conditions and periods of water deficit, such as the genera *Tillandsia* (Bromeliaceae) and *Pleopeltis*

(Polypodiaceae) indicate that environmental factors are the main limiting factors for the development of epiphytes in the area (BREIER, 2005; BATAGHIN et al., 2008). Although cerrado areas present limiting conditions for the development of epiphytes, especially as they present adverse climatic conditions for epiphytism (LAUBE; ZOTZ, 2003), there seems to be some adaptive success among some families, such as Bromeliaceae and Polypodiaceae, either due to the presence of water storage mechanisms or due to their resistance to desiccation, especially if anthropic interference is not intense.

The Polypodiaceae family, cited by Breier (2005) and Ishara et al. (2008) as the richest in Cerrado areas, was among the richest families at this site. This family was joined by Bromeliaceae and Orchidaceae, observed as the richest in the study by Bataghin et al. (2012b). These three families had six species each and accounted for 75% of the site's species.

The characteristic holoepiphytes were predominant, accounting for almost 84% of the species, followed by primary hemiepiphytes (8%), facultative holoepiphytes and accidental holoepiphytes, both with 4% each, with no secondary hemiepiphytes recorded. Anemochory was predominant as a dispersal syndrome, being present in 17 species (71%) while zoochory was recorded in seven species (29%). The presence of a greater number of anemochorous species is expected in vascular epiphyte communities (Benzing 1987).

In the quantitative analysis, 17 species were recorded, with two Polypodiaceae being the most prominent species at Sitio Réplica II (Table 10).

Table 10 - Vascular epiphytes from Sitio Réplica II in the Downstream Area of the Sorocaba Médio Tietè hydrographic basin, classified according to their epiphytic importance value - nr: absolute number of occurrences in the strata; far: absolute frequency in the strata; ni: absolute number of occurrences in the forophytic individuals; fai: absolute frequency in the forophytic individuals; vt (total value): sum of the abundance estimates; vie: epiphytic importance value; nota: average score obtained.

Species	nr	far	ni	fai	vt	vie	note
Pleopeltis pleopeltifolia	265	49.1	79	87.8	439	37.71	1.66
Microgramma squamulosa	128	23.7	34	37.8	234	20.10	1.83
Tillandsia recurvata	112	20.7	34	37.8	163	14.00	1.46
Tillandsia tricholepis	57	10.6	19	21.1	85	7.30	1.49
Pleopeltis squalida	26	4.8	8	8.9	52	4.47	2.00
Tillandsia stricta	34	6.3	13	14.4	52	4.47	1.53
Aechmea bromeliifolia	22	4.1	12	13.3	49	4.21	2.23

Keep going...

Table 10 - Contents...

Species	nr	far	ni	fai	vt	vie	note
Lepismium lumbricoides	13	2.4	7	7.8	21	1.80	1.62
Acanthostachys strobilacea	9	1.7	5	5.6	18	1.55	2.00
Microgramma tecta	9	1.7	3	3.3	13	1.12	1.44
Epiphyllum phyllanthus	7	1.3	4	4.4	12	1.03	1.71
Pleopeltis hirsutissima	5	0.9	2	2.2	9	0.77	1.80
Tillandsia funckiana	5	0.9	3	3.3	5	0.43	1.00
Vittaria lineata	3	0.6	2	2.2	5	0.43	1.67
Serpocaulon latipes	1	0.2	1	1.1	3	0.26	3.00
Oeceoclades maculata	1	0.2	1	1.1	2	0.17	2.00
Polystachya foliosa	1	0.2	1	1.1	2	0.17	2.00

Pleopeltis pleopeltifolia (Figure 10) had a value of epiphytic importance (VIE) of 37.71 and an average score of 1.66, occurring in almost 88% of the forophytes and 49.1% of the strata, making it the most important species at this site. *Microgramma squamulosa* had a VIE of 20.10, an average score of 1.83 and was recorded in 37.8% of the forophytes and 23.7% of the strata.

Figura 10:   Detail of the species *Pleopeltis pleopeltifolia* (Raddi) Alston (Polypodiaceae), its predominance is characteristic of forest areas of low structural complexity.

Two species of Bromeliaceae are also worth mentioning: Tillandsia recurvata with a VIE of 14.0 and an average score of 1.46 and Tillandsia tricholepis with a VIE of 7.3 and an average score of 1.49. These species contribute to the Bromeliaceae family being the second most important in the site, reaching 31.96% of the epiphytic importance value. The Polypodiaceae family, the most important, was responsible for 64.43% of the VIE. The Orchidaceae family, despite being one of the richest in this site, showed a low relative abundance with a VIE of 0.34, a possible indication of the family's low adaptation to cerrado areas or the occurrence of smaller populations.

The concentration of richness and abundance in the Polypodiaceae and Bromeliaceae families is characteristic of areas with adverse climatic conditions for epiphytism, either due to the natural characteristics of the vegetation (Breier 2005) or the interference of anthropogenic activities (DETTKE et al., 2008; BATAGHIN et al., 2010; BATAGHIN et al., 2012a).

Analysis of the forophytic strata revealed that the base of the canopy (Figure 11) was the most abundant with an abundance value (AV) of 385; the inner canopy was the second most abundant stratum with an AV of 296, followed by the high stem with an AV of 221. The

51

middle stem, outer canopy and low stem had abundance values of 107, 99 and 57 respectively.

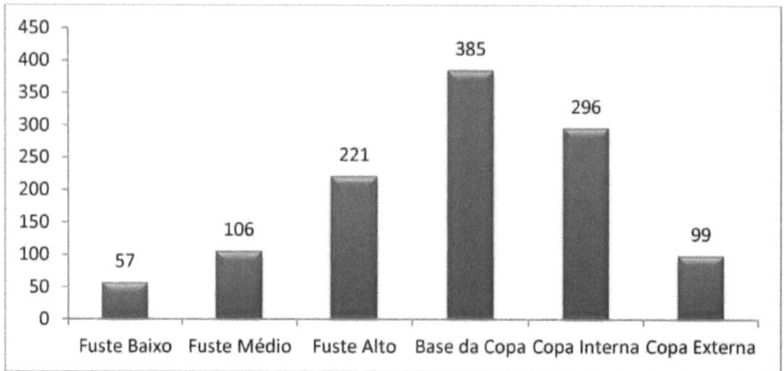

Figura 11: Distribution of the abundance of epiphytic vascular species among the forophytic strata at Sitio Réplica II in the Downstream Area of the Sorocaba/Médio Tietê hydrographic basin.

Analysis of the distribution of abundances revealed only two significant differences involving the low stem and the base of the canopy and the low stem and the inner canopy

(Table 11). In terms of richness, the outer canopy was significantly different from all the other strata, except for the low stem.

Table 11: Analysis of the distribution of the richness and abundance of vascular epiphytes among the forophytic strata at Sitio Réplica II in the Downstream Area of the Sorocaba/Médio Tietê hydrographic basin.

^^^-^^^Richness Abundance"----^	Beam Bass	Beam Medium	Beam High	Base of Cup	Cup Internal	Cup External
Low stem		0,7435	0,2881	0,1439	0,2881	0,0514
Medium stem	0,184		0,4716	0,26	0,4716	**0,0182**
High stem	0,079	0,174		0,6941	0,9813	**0,0025**
World Cup base	**0,027**	0,054	0,205		0,6941	**0,0008**
Internal Cup	**0,038**	0,087	0,331	0,335		**0,0025**
External Cup	0,271	0,465	0,173	0,055	0,089	

Although there is a concentration of epiphyte abundance at Sitio Réplica II in the intermediate strata, especially at the base of the canopy and in the inner canopy, where the presence of branch insertion points that can favor the installation of epiphytes is striking (FREIBERG, 1996; NIEDER et al., 1999), the importance of microclimatic factors in this result is clear. The absence of significant differences between the strata indicates that climatic factors which are adverse to epiphytic development penetrate the forest more intensely, reaching almost the limit of the soil, causing the vascular epiphytic community to show a more homogeneous vertical distribution, especially in terms of richness, on the forophyte. However, the limitation of species that grow in the outer canopy, where they need to withstand extreme conditions in terms of humidity and light, is clear.

In addition, the presence of similar species between the low stem and the outer layer, together with the concentration of abundance in a few species (Table 10) and typical of impacted areas (DETTKE et al., 2008; BATAGHIN et al., 2012a) serve as an indication that this forest fragment shows some degree of degradation. The vertical distribution of epiphytes seems to be a useful tool for characterizing the conservation status of forest remnants.

Analysis of the vascular epiphytes at Sitio Réplica III in the downstream area of the Sorocaba/Médio Tietê basin

The survey of vascular epiphytes at Sitio Réplica III, in the downstream area of the Sorocaba/Médio Tiete river basin, was carried out on a private rural property called Fazenda Sao Francisco do Tietê, in the municipality of Anhembi - SP. The forest remnant covers around 500 ha and is located between UTM coordinates 788.448 and 7.490.010 in zone 22 South, and is characterized by being a stretch of ecotone between the Semideciduous Seasonal Forest and the Cerrado (Figure 12). Although signs of anthropogenic influence were found at Sitio Réplica III, such as the presence of trails, the forest area can be characterized as well preserved, both due to the (abundant) presence of larger trees and the structure of the forest, especially the presence of a denser understorey. In addition, an extensive flat area like the one at this site is rarely found with vegetation in this region of the Sorocaba/Médio Tietê basin. The surrounding areas are characterized by anthropic use, such as agricultural cultivation (manioc) and cattle breeding.

Figura 12: Forest fragment of Sitio Réplica III in the Downstream Area: A - external view of the area; B - aspect of the understorey.

Sitio Réplica III was the area with the greatest diversity of epiphytic species in the Downstream Area of the Sorocaba/Médio Tietê basin, with 35 species recorded, belonging to 20 genera and seven families (Table 12). The Shannon index for the area was H' = 3.119, the equability (J) was 0.877 and the Margalef richness (d) was 4.795.

Table 12 - List of vascular epiphyte species found at Sitio Réplica III in the downstream area (Fazenda Sao Luiz) of the Sorocaba Médio Tietê hydrographic basin and their respective Ecological Categories (EC) - HLC: Characteristic holoepiphyte; HLF: Facultative holoepiphyte; HLA: Accidental holoepiphyte; HMP: Primary hemiepiphyte; HMS: Secondary hemiepiphyte. Dispersal form (Disp.) - Zo: Zoochoric; An: Anemochoric. Reg: HUFSCar herbarium registration number (Im: Digital image).

N	Family	Species	EC	Disp.	Reg.
1	Araceae	Philodendron bipinnatifidum Schott	HMP	Zo	Im
2	Bromeliaceae	Acanthostachys strobilacea (Schult. f.) Klotzsch	HLC	Zo	8431
3	Bromeliaceae	Tillandsia funckiana Baker	HLC	An	8458
4	Bromeliaceae	Tillandsia recurvata (L.) L.	HLC	An	8429
5	Bromeliaceae	Tillandsia tricholepis Baker	HLC	An	8430
6	Bromeliaceae	Tillandsia usneoides (L.) L.	HLC	An	8460
7	Bromeliaceae	Vriesea bituminosa Wawra	HLC	An	8515
8	Cactaceae	Epiphyllum phyllanthus (L.) Haw.	HLC	Zo	8474
9	Cactaceae	Lepismium cruciforme (Vell.) Miq.	HLC	Zo	8388
10	Cactaceae	Lepismium lumbricoides (Lem.) Barthlott	HLC	Zo	8386
11	Cactaceae	Rhipsalis cereuscula Haw.	HLC	Zo	8383
12	Cactaceae	Rhipsalis teres (Vell.) Steud.	HLC	Zo	8384
13	Commelinaceae	Tradescantia albiflora Kunth	HLA	Zo	8391
14	Orchidaceae	Acianthera nemorosa (Barb. Rodr.) F. Barros	HLC	An	8403
15	Orchidaceae	Acianthera saundersiana (Rchb.f.) Pridgeon & M.W.Chase	HLC	An	8401
16	Orchidaceae	Baptistonia lietzei (Regel) Chiron & V.P.Castro	HLC	An	8407
17	Orchidaceae	Bulbophyllum chloroglossum Rchb.f. & Warm.	HLC	An	8408
18	Orchidaceae	Epidendrum rigidum Jacq.	HLC	An	8454
19	Orchidaceae	Octomeria crassifolia Lindl.	HLC	An	8472
20	Orchidaceae	Octomeria gracilis Lodd. ex Lindl.	HLC	An	8440
21	Orchidaceae	Octomeria palmyrabellae Barb. Rodr.	HLC	An	8442
22	Orchidaceae	Ornithocephalus myrticola Lindl.	HLC	An	8443
23	Orchidaceae	Rodriguezia decora Rchb. f.	HLC	An	8441
24	Piperaceae	Peperomia glabella (Sw.) A. Dietr.	HMP	Zo	8475
25	Piperaceae	Peperomia rotundifolia (L.) Kunth	HLC	Zo	8449
26	Piperaceae	Peperomia tetraphylla (G. Forst.) Hook. & Arn.	HLC	Zo	8447
27	Piperaceae	Peperomia trineura Miq.	HLC	Zo	8446
28	Polypodiaceae	Campyloneurum sp.	HLC	An	8517
29	Polypodiaceae	Microgramma persicariifolia (Schrad.) C. Presl	HLC	An	8520
30	Polypodiaceae	Microgramma squamulosa (Kaulf.) de la Sota	HLC	An	8534

31	Polypodiaceae	Microgramma tecta (Kaulf.) Alston	HLC	An	8524
32	Polypodiaceae	Pleopeltis hirsutissima (Raddi) de la Sota	HLC	An	8521
33	Polypodiaceae	Pleopeltis pleopeltifolia (Raddi) Alston	HLC	An	8525
34	Polypodiaceae	Pleopeltis squalida (Vell.) de la Sota	HLC	An	8436
35	Polypodiaceae	Serpocaulon latipes (Langsd. & L. Fisch.) A.R. Sm.	HLF	An	8423

The richness of Sitio Réplica III can be compared to that of studies carried out in Semideciduous Seasonal Forest, being higher than the data obtained by Aguiar et al. (1981), who sampled 17 species; by Breier (2005), with 25 species, and by Bataghin et al. (2010), with 21 species and similar to the data presented by Dislich and Mantovani (1998), with 34 species; by Borgo et al. (2002), with 32 species and by Dettke et al. (2008), with 29 species. Despite this, the richness was lower than the results of Rogalski and Zanin (2003) who found 70 species, Giongo and Waechter (2004) who sampled 57 species and Cervi and Borgo (2007) who found 56 species. It is important to note that these last three studies were carried out in forests in southern Brazil, which differ fundamentally in terms of the greater supply of water resources to epiphytes compared to the areas downstream of the watershed studied here. When compared to surveys carried out in cerrado areas, the richness of the area is higher than the studies by Breier (2005) who observed 16 species, Ishara et al. (2008) - seven species and Joanitti et al. (2010) - 16 species, and similar to the data obtained by Bataghin et al. (2012b) who found 29 species.

Sitio Rèplica III had the highest diversity of vascular epiphytes in the Downstream Area, and the better conservation of the forest area studied certainly has an influence on these numbers. Several studies have reported that changes in the landscape have a negative influence on the diversity and abundance of vascular epiphytes (ENGWALD et al., 2000; BONNET; QUEIROZ, 2000; BARTHLOTT et al., 2001, BATAGHIN et al., 2008, DETTKE et al.., 2008), however, forest remnants that are not protected by UCs should not be ignored or left aside in conservation actions, as it seems that forest fragmentation may have isolated or even restricted the epiphytic community in these forest fragments.

The richest families at Sitio Rèplica III were Orchidaceae with 10 species and Polypodiaceae with eight species. The Bromeliaceae and Cactaceae families had six and five species, respectively. These were followed by Piperaceae, which had four species, and Araceae and Commelinaceae with one species each. The characteristic holoepiphytes were dominant at the site, accounting for 31 species (88%), followed by facultative holoepiphytes, 2 spp. (6%), accidental holoepiphytes and primary hemiepiphytes with one species each (3%) and no secondary hemiepiphytes were recorded. This same pattern of classification into ecological categories was observed at both the Core Site and the other sites in this area of the river basin.

With regard to the dispersal syndrome, 23 species were classified as anemochoric (66%) and 12 species as zoochoric, following the pattern found in other studies (BENZING, 1987; BREIER, 2005).

All 35 species found in the floristic survey were recorded in the quantitative analysis of Sitio Rèplica III (Table 13).

Table 13 - Vascular epiphytes from Replica Site III in the Downstream Area of the Sorocaba-Mèdio Tietê river basin, classified according to epiphytic importance value - nr: absolute number of occurrences in the strata; far: absolute frequency in the strata; ni: absolute number of occurrences in forophytic individuals; fai: absolute frequency in forophytic individuals; vt (total value): sum of abundance estimates; vie: epiphytic importance value; nota: average score obtained.

Species	nr	far	ni	fai	vt	vie	note
Microgramma squamulosa	60	11.1	20	22.2	114	9.50	1.90
Rhipsalis teres	59	10.9	22	24.4	106	8.83	1.80
Lepismium lumbricoides	56	10.4	23	25.6	101	8.42	1.80
Peperomia rotundifolia	50	9.3	26	28.9	94	7.83	1.88
Pleopeltis pleopeltifolia	59	10.9	19	21.1	83	6.92	1.41
Tillandsia recurvata	57	10.6	22	24.4	74	6.17	1.30
Tillandsia tricholepis	54	10.0	21	23.3	74	6.17	1.37
Microgramma persicariifolia	40	7.4	14	15.6	65	5.42	1.63
Microgramma tecta	35	6.5	14	15.6	49	4.08	1.40
Tradescantia albiflora	27	5.0	13	14.4	45	3.75	1.67
Lepismium cruciforme	27	5.0	13	14.4	41	3.42	1.52
Pleopeltis squalida	22	4.1	7	7.8	39	3.25	1.77
Tillandsia funckiana	22	4.1	9	10.0	36	3.00	1.64
Peperomia trineura	18	3.3	11	12.2	34	2.83	1.89
Pleopeltis hirsutissima	18	3.3	8	8.9	33	2.75	1.83
Philodendron bipinnatifidum	14	2.6	7	7.8	30	2.50	2.14
Rodriguezia decorates	12	2.2	4	4.4	22	1.83	1.83
Baptistonia lietzei	10	1.9	5	5.6	18	1.50	1.80
Octomeria crassifolia	8	1.5	5	5.6	16	1.33	2.00
Tillandsia usneoides	9	1.7	4	4.4	15	1.25	1.67
Vriesea bituminosa	5	0.9	3	3.3	12	1.00	2.40
Bulbophyllum chloroglossum	4	0.7	2	2.2	12	1.00	3.00
Acianthera saundersiana	4	0.7	2	2.2	11	0.92	2.75
Acanthostachys strobilacea	5	0.9	3	3.3	10	0.83	2.00

Epiphyllum phyllanthus	6	1.1	3	3.3	10	0.83	1.67
Octomeria gracilis	4	0.7	2	2.2	10	0.83	2.50
Ornithocephalus myrticola	5	0.9	2	2.2	9	0.75	1.80
Serpocaulon latipes	7	1.3	3	3.3	9	0.75	1.29
Peperomia tetraphylla	4	0.7	2	2.2	6	0.50	1.50
Rhipsalis cereuscula	2	0.4	1	1.1	5	0.42	2.50
Campyloneurum sp.	2	0.4	2	2.2	4	0.33	2.00
Octomeria palmyrabellae	2	0.4	1	1.1	4	0.33	2.00
Acianthera nemorosa	2	0.4	1	1.1	3	0.25	1.50

Table 13 - Continuation...

Species	nr	far	ni	fai	vt	vie	note
Epidendrum rigidum	2	0.4	1	1.1	3	0.25	1.50
Peperomia glabella	2	0.4	1	1.1	3	0.25	1.50

Micrrogramma squamulosa (Polypodiaceae) had an epiphytic importance value (EVI) of 9.50 and an average score of 1.90, being present in 22% of the forophytes and 11% of the strata, making it the most important species in this site. *Rhipsalis teres* (Cactaceae) was the second most important species with a VIE of 8.83 and an average score of 1.80, as well as being present in 24.4% of the forophytes and 10.9% of the strata. The third most important species at this site was *Lepismium lumbricoides* (Cactaceae), with a VIE of 8.42 and a score of 1.80, and it was recorded in over 25.6% of the forophytes and 10.4% of the strata. Among the Orchidaceae, *Rodriguezia decora* (Figure 13) had the highest importance value VIE=1.83, although it was recorded in only four forophytes.

Figura 13:   Detail of the species *Rodriguezia decora* Rchb. f. (Orchidaceae), the most

abundant of the orchids at Sitio Réplica III in the Downstream Area.

At the Replica III site in the Downstream Area, the Polypodiaceae family was the most important, with 33% of the epiphytic importance value. The Cactaceae family was the second most important with a total VIE of 21.92, followed by Bromeliaceae with a VIE of 18.42 and Piperaceae with a VIE of 11.42. The Orchidaceae family deserves attention, because although it has the highest number of species for the site (10 spp.), it had a low importance value (VIE = 8.9). Although orchids are more sensitive to environmental variations (Ditt 2002), which reduces their numbers in many forest remnants, for the Orchidaceae family, there is a tendency for small populations to occur, even in well-preserved forest fragments.

Although the Polypodiaceae, Cactaceae and Bromeliaceae - significant families in areas under anthropogenic pressure (DETTKE et al., 2008), are predominant at the site, the greater diversity of species in relation to the other sites studied in this area of the basin indicates an environment favorable to the establishment and development of vascular epiphytes. It is important to note that there is no dominance of species in relation to the distribution of epiphytic importance values at this site (Table 13). This distribution can be considered to reflect a dynamic balance in the epiphytic community, which has evolved over a long period of time, indicative of the good state of conservation of the forest area at this site. Normally, areas that have been altered or are in a successional stage are dominated by a few species (BATAGHIN et al., 2012a; BARTHLOTT et al., 2001).

With regard to the vertical distribution of epiphytes, the base of the canopy (Figure 14) was the forophytic stratum with the highest abundance (VA = 381), followed by the inner canopy, the second most abundant stratum with VA of 288 and the high stem with VA of 179. The middle canopy had an abundance value of 144, while the low canopy and outer canopy had abundance values of 131 and 77, respectively.

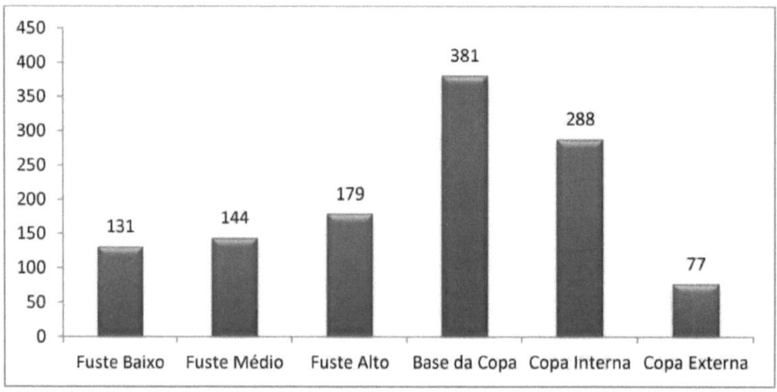

Figura 14: Distribution of the abundance of epiphytic vascular species among the forophytic strata at Sitio Règlica III in the Downstream Area of the Sorocaba/Médio Tietê hydrographic basin.

The analysis applied to the distribution of vascular epiphytes in the forophytic strata of Sitio Réplica III (Table 14) showed that, in terms of abundance, the base of the canopy and the inner canopy were similar to each other and different from the other strata, except for the inner canopy, which showed no significant difference from the high stem, the latter also showing a significant difference from the outer canopy. In terms of richness, the low stem and the outer canopy showed different communities from all the other strata, including each other. In addition, the base of the canopy was significantly different from all the strata, except the inner canopy.

Table 14: Analysis of the distribution of the richness and abundance of vascular epiphytes among the forophytic strata at Sitio Réplica III in the Downstream Area of the Sorocaba/Médio Tietê basin.

^^^^Richness Abundance^-^^	Beam Bass	Beam Medium	Beam High	World Cup base	Cup Internal	Cup External
Low stem		**0,0329**	**0,0088**	**7,03E-06**	**0,0007**	**0,03627**
Medium stem	0,427		0,6143	**0,009472**	0,1846	**4,34E-05**
High stem	0,24	0,276		**0,03339**	0,4109	**5,84E-06**
World Cup base	**0,004**	**0,004**	**0,009**		0,1781	**6,15E-10**
Internal Cup	**0,041**	**0,042**	0,087	0,181		**1,82E-07**
External Cup	0,227	0,148	**0,047**	**4,23E-04**	**0,007**	

The concentration of epiphyte abundance in the intermediate strata, especially at the base of the canopy and in the inner canopy at Sitio Réplica III, was expected and has been common in studies carried out in dry forests (BREIER, 2005; BATAGHIN et al., 2012b). However, the different distribution of richness in the low stem, high stem and middle part of the forophyte indicates the occupation of specific niches by different species on the host tree.

At this site, attention is drawn to the greater number of individuals in the lower strata compared to those observed in the outer canopy (Figure 14). In addition, the community (richness) that develops in the low stem was significantly different from that observed in the outer canopy (which is generally made up of species with greater environmental plasticity and appear in all forophyte strata), which is an indication of the good state of conservation of the forest remnant, since in the other sites in the Downstream Area (more impacted) a low abundance was recorded in the lower strata and the richness of the low stem and outer canopy were similar.

In addition, in this site of better conserved forest, there are indications that the mechanisms that maintain epiphytic diversity go beyond climatic factors (such as humidity and luminosity) to more specific factors, such as competition between species.

Analysis of the vascular epiphytes of the Qualitative Sites (QI, QII, QIII) in the Downstream Area of the Sorocaba/Médio Tietê basin

The survey of vascular epiphytes in Qualitative Sites QI, QII and QIII in the downstream area of the Sorocaba/Médio Tiete river basin was carried out in three different areas. Qualitative Site I (QI) was located in an area of approximately 125 ha, between UTM coordinates 766.618 and 7.474.496 in zone 22 South and the vegetation can be characterized as an ecotone region between the Semideciduous Seasonal Forest and the Cerrado. The forest remnant of Sitio Qualitativo II (QII) has around 115 ha and is located between UTM coordinates 769.818 and 7.484.139 in zone 22 South, and is characterized as a stretch of Cerrado. Qualitative Site III (QIII), located between UTM coordinates 774.135 and 7.485.128 in zone 22 South, is a 154 ha forest fragment located inside a farm that grows orange, and can be characterized as a mixed area between Cerrado and Cerrado *stricto sensu* (the survey was carried out only in the Cerrado area). In all three qualitative sites, signs of anthropic influence were found, such as the presence of trails, traces of fishing equipment near the waterbeds and traps for the illegal capture of wild animals; it should also be noted that in Site QII, a wooden frame was found that had been built to serve as a "waiting place" for illegal hunters (Figure 15). The areas surrounding the qualitative sites can be characterized by anthropic use, such as agricultural cultivation (sugar cane, oranges and rice) and cattle breeding.

Figura 15: Traps for hunting wild animals: A - jirau-type trap (wooden frame used by hunters as a "wait"); B - bait trap.

At the three Qualitative Sites (QI, QII, QIII) in the Downstream Area of the Sorocaba/Médio

Tietê river basin, 29 species were recorded, belonging to 17 genera and six families (Table 15).

Table 15 - List of the vascular epiphyte species found at the Qualitative Sites in the Downstream Area of the Sorocaba Mèdio Tietê river basin and their respective Ecological Categories (EC) - HLC: Characteristic Holoepiphyte; HLF: Facultative Holoepiphyte; HLA: Accidental Holoepiphyte; HMP: Primary Hemiepiphyte; HMS: Secondary Hemiepiphyte. Dispersal form (Disp.) - Zo: Zoochoric; An: Anemochoric. Q: Qualitative Survey Site (1, 2, 3).

N	Family	Species	EC	Disp.	Sitio
1	Araceae	Philodendron bipinnatifidum Schott	HMS	Zo	Q1;
2	Bromeliaceae	Acanthostachys strobilacea (Schult. f.) Klotzsch	HLC	Zo	Q2; Q3;
3		Aechmea apocalyptica Reitz	HLF	Zo	Q2;
4		Aechmea bromeliifolia (Rudge) Baker	HLC	Zo	Q1; Q2; Q3
5		Aechmea distichantha Lem.	HLC	Zo	Q1; Q2;
6		Billbergia amoena (Lodd.) Lindl.	HLC	Zo	Q2;
7		Tillandsia funckiana Baker	HLC	An	Q2;
8		Tillandsia recurvata (L.) L.	HLC	An	Q1; Q2; Q3
9		Tillandsia stricta Sol. ex Sims	HLC	An	Q1; Q2; Q3
10		Tillandsia tricholepis Baker	HLC	An	Q1; Q3;
11		Vriesea procera (Mart. ex Schult. & Schult.f.) Wittm.	HLC	An	Q2;
12	Cactaceae	Cereus alacriportanus Pfeiff.	HLA	Zo	Q1;
13		Epiphyllum phyllanthus (L.) Haw.	HLC	Zo	Q1; Q2;
14		Lepismium cruciforme (Vell.) Miq.	HLC	Zo	Q1;
15		Lepismium lumbricoides (Lem.) Barthlott	HLC	Zo	Q1;
16		Rhipsalis cereuscula Haw.	HLC	Zo	Q1;
17		Rhipsalis teres (Vell.) Steud.	HLC	Zo	Q1;

Table 15 - Contents...

N	Family	Species	EC	Disp.	Sitio
18	Cactaceae	*Rhipsalis trigona* Pfeiff.	HLC	Zo	Q2;
19	Orchidaceae	*Acianthera recurva* (Lindl.) Pridgeon & M.W. Chase	HLC	An	Q2;
20		*Polystachya estrellensis* Rchb. f.	HLC	An	Q1;
21	Piperaceae	*Peperomia rotundifolia* (L.) Kunth	HLC	Zo	Q1; Q2;
22	Polypodiaceae	*Microgramma tecta* (Kaulf.) Alston	HLC	An	Q2;
23		*Microgramma squamulosa* (Kaulf.) de la Sota	HLC	An	Q1; Q2;
24	Polypodiaceae	*Pecluma filicula* (Kaulf.) M.G. Price	HLC	An	Q2;
25		*Pleopeltis astrolepis* (Liebm.) E. Fourn.	HLC	An	Q2;
26		*Pleopeltis hirsutissima* (Raddi) de la Sota	HLC	An	Q1 ;Q2; Q3
27		*Pleopeltis pleopeltifolia* (Raddi) Alston	HLC	An	Q1; Q2; Q3
28		*Pleopeltis squalida* (Vell.) de la Sota	HLC	An	Q1; Q2;
29		*Serpocaulon latipes* (Langsd. & Fisch.) A.R.Sm	HLF	An	Q2; Q3;

At Sitio QI, 18 species belonging to 11 genera and six families were recorded. Characteristic holoepiphytes were predominant at this site (16 species), followed by accidental holoepiphytes and secondary hemiepiphytes (one species each). No facultative holoepiphytes or primary hemiepiphytes were recorded at Sitio QI. The Cactaceae family was the richest at this site with six species, followed by Bromeliaceae and Polypodiaceae with five and four species, respectively. Araceae, Orchidaceae and Piperaceae had one species each.

Site QII was the richest among the qualitative sites in the Downstream Area, with 21 species belonging to 13 genera and five families. Characteristic holoepiphytes accounted for 18 species, followed by accidental holoepiphytes with three. Bromeliaceae, with 10 species (Figure 16), and Polypodiaceae, with eight species, were dominant at this site. Cactaceae accounted for two species and the Orchidaceae and Piperaceae families had one species each.

Site QIII was the least diverse of all the sites in the Downstream Area, with only eight species of vascular epiphytes, belonging to five genera and only two families (Bromeliaceae - five species and Polypodiaceae - three species).

Figura 16:   Detail of Aechmea bromeliifolia (Rudge) Baker (Bromeliaceae), found in the three Qualitative Sites of the Downstream Area of the hydrographic basin.

The richness of the Qualitative Sites (QI, QII, QIII), analyzed together here, is low when compared to the works by Rogalski and Zanin (2003) who found 70 species, Giongo and Waechter (2004) who sampled 57 species and Cervi and Borgo (2007) who found 56 species. However, it is similar to the studies carried out in Semideciduous Seasonal Forest by Aguiar et al. (1981), who sampled 17 species; by Dislich and Mantovani (1998), with 34 species; by Borgo et al. (2002), with 32 species; by Breier (2005), with 25 species, by Dettke et al. (2008), with 29 species and by Bataghin et al. (2010), with 21 species. It can also be considered superior to the surveys carried out in cerrado areas, such as the studies by Breier

(2005) which found 16 species, Ishara et al. (2008) - seven species and Joanitti et al. (2010) - 16 species, and similar to the data obtained by Bataghin et al. (2012b) which found 29 species. Analyzing the isolated case of Sitio QIII, which showed only eight species, it can be said that this result is not unusual, given that Ishara et al. (2008), studying a forest environment similar to that of this site, found seven epiphytic species.

The richest families in the Qualitative Sites in the downstream area (QI, QII, QIII) were Bromeliaceae with 10 species, Polypodiaceae with eight species and Cactaceae with seven species. The Orchidaceae family had two species and Araceae and Piperaceae had one species each. In general, the characteristic holoepiphytes were dominant at the site, accounting for 24 species (83%), followed by facultative holoepiphytes with three species (10%), and accidental holoepiphytes and secondary hemiepiphytes with one species each (3.5%). No primary hemiepiphytes were recorded at the qualitative sites in the Downstream Area. Of the total number of species, 15 (52%) showed anemochoric dispersal and 14 (48%) zoochoric dispersal. This result seems to repeat the pattern observed at the Réplica I site in the Downstream Area, where the anemochoric and zoochoric dispersal syndromes had a certain equivalence, which was not expected, since anemochory is predominant for this syntomy (BENZING, 1987). This provides an indication that small forest fragments isolated in the agricultural landscape, anthropically altered or of low structural complexity, tend to have a higher number of epiphytic species with a zoochoric dispersal syndrome. This occurs for two reasons: i - the greater distance between forest fragments acts as a barrier to anemochorically-dispersed epiphytic species, reducing their number; ii - the reduction in the structural complexity of the forest fragment excludes anemochorically-dispersed species that are more sensitive in microclimatic terms, such as orchids.

Analysis of vascular epiphytes in the Ecotone between the Semideciduous Seasonal Forest and the Cerrado in the Downstream Area of the Sorocaba/Médio Tietê Watershed

The statistical analysis applied to the distribution of vascular epiphyte abundances between the sites showed no significant differences between the Core Site and its replicas. The values obtained were as follows: the analysis between the Core Site and Replica Site I showed t = - 0.430 and p = 0.334; between the Core Sites and Replica II, t = - 0.027 and p = 0.489; and between the Core Sites and Replica III, t = -0.142 and p = 0.444. The absence of a significant difference between the abundances of the Core Sites and their replicas may indicate that the vascular epiphytic community that develops in the Semideciduous Seasonal Forest/Cerrado Ecotone of the Downstream Area in the Sorocaba/Médio Tietê watershed can be considered similar in terms of the number of individuals in the different sites studied.

The same statistical analysis applied to the presence/absence of species showed that the Core Site was different from all its Replicas, except for Replica Site II. The analysis yielded the following results: between Core Site and Replica I: $t = 4.494$ and $p = 0.001$, between Core Site and Replica II: $t = 0.237$ and $p = 0.407$, between Core Site and Replica III: $t = -2.333$ and $p = 0.011$, between Sitio Core and QI: $t = 1.967$ and $p = 0.028$, between Sitio Core and QII: $t = 4.192$ and $p = 0.0001$, and between Sitio Core and QIII: $t = 5.762$ and $p = 0.0001$. However, when comparing the diversity of the Core Site (a Conservation Unit - 25 species) with the total diversity of the other sites (unprotected forest fragments in the surrounding area - 51 species), a significant difference was found with values of $t = -6.008$ and $p = 0.0001$. It is also important to note that 31 species from the downstream area were recorded only in unprotected forest fragments, highlighting the vulnerability of a large portion of the epiphytic vascular community. The analysis reflects the diversity of species found in the area as a whole and highlights the importance of conserving forest fragments for vascular epiphytic diversity, whether these fragments are protected by Conservation Units or located in private areas.

The Jaccard Similarity Index, which can be seen in Figure 17, showed similarities of: 33% between Sitio Core and Sitio Réplica I; 29% between Sitio Core and Sitio Réplica II; 43% between Sitio Core and Sitio Réplica III; 42% between Sitio Core and Sitio QI; 25% between Sitio Core and Sitio QII; and 18% between Sitio Core and Sitio QIII. The greatest floristic similarity occurred between Sitios Réplica I and Sitio QI, which shared 50% of the species. This low level of similarity between the sites corroborates the idea that the vascular epiphytic community found in the Downstream Area of the Sorocaba/Médio Tietê basin is heterogeneous, revealing the importance of forest fragments not protected by UCs in the composition of the vascular epiphytic flora.

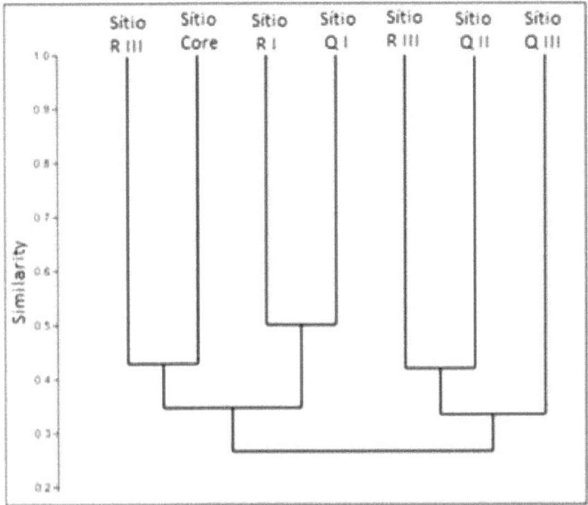

Figura 17:   Dendrogram (UPGMA) of Jaccard's similarity (presence/absence) between sites in the Ecotone between Semideciduous Seasonal Forest and Cerrado in the downstream area of the Sorocaba/Médio Tietê basin.

Another interesting aspect is the similarity between the vascular epiphytes sampled in the Downstream Area of the Sorocaba/Médio Tietê basin and other studies carried out on vascular epiphytes in different forest formations, both in areas of Semideciduous Seasonal Forest and in Cerrado areas in Brazil (Table 16).

Table 16 - Jaccard's similarity index (J) between the vascular epiphytic species of the Downstream Area of the Sorocaba/Médio Tietê basin and other studies of vascular epiphytes carried out in Semideciduous Seasonal Forest and Cerrado in Brazil. FES: Semideciduous Seasonal Forest; FOM: Mixed Ombrophilous Forest; CER: Cerrado.

Source	Location (Phytophysiognomy)	Species	Families	Jaccard (J)
Aguiar et al. 1981	Montenegro - RS (FES)	17	5	0,163
Dislich and Mantovani 1998	Sao Paulo - SP (FES)	34	9	0,193
Borgo et al. 2002	Fenix - PR (FES)	32	10	0,172
Rogalski and Zanin 2003	Marcelino Ramos - RS (FES)	70	8	0,067
Giongo and Waechter 2004	Eldorado do Sul - RS (FES)	57	13	0,246
Breier 2005	Assis - SP (FES)	25	9	0,229
Cervi and Borgo 2007	Foz do Iguaçù - PR (FES)	56	13	0,139
Dettke et al. 2008	Maringa - PR (FES)	29	8	0,182
Menini-Neto et al. 2009	Barroso - MG (FES)	41	5	0,061
Menini-Neto et al. 2009	Descoberto - MG (FES)	59	10	0,062
Alves and Kolbek 2009	Tiradentes/Prados - MG (FES)	61	12	0,148
Geraldino et al. 2010	Campo Mourao - PR (FES/FOM)	61	12	0,232

Bataghin et al 2010	Iperó - SP (FES)	21	5	0,783
Bonnet et al. 2011	Tibagi River Basin - PR (FES)	60	13	0,262
Linsingen et al. 2006	Jaguariaiva - PR (FES/CER)	16	3	0,088
Breier 2005	Assis - SP (CER)	16	5	0,167
Ishara et al. 2008	Botucatu - SP (CER)	7	3	0,104
Bataghin et al 2012b	Luiz Antonio - SP (CER)	29	7	0,432

The low level of similarity observed between the studies and the results obtained for the Downstream Area may be associated with the great sensitivity of the epiphytic community to climatic and microclimatic variations (BENZING, 1990, 1995), given the environmental changes suffered by the forests in the Sorocaba/Médio Tietê basin (ZERO REPORT, 2005), or the climatic differences between the sites, since studies carried out in geographic regions close to each other tend to show greater floristic similarity (MENINI-NETO et al., 2009). Environmental changes, especially forest fragmentation and loss, are the main causes of microclimatic changes and variations in forests. Dettke et al. (2008) and Barthlott et al. (2001) point out that the occurrence of epiphytic species is related to the integrity of the forest and, consequently, to favorable climatic conditions. However, the variations caused by the natural distribution of species in different forest physiognomies, and especially the occurrence of different species according to Iatitudinal variations (WAECHTER, 1998), is one possibility for the low similarity between the epiphyte species that occur in the area studied and the species observed by different authors (Table 16). The irregular geographical distribution pointed out by Kersten (2006) may be another important factor behind the low similarity observed. It is interesting to note that the greatest similarities occur with different forest formations in nearby areas (FES / Iperó-SP - basin area) or relatively close to the sites studied (Cerrado / Luiz Antonio-SP). Although there is a general tendency for epiphytic diversity to decrease from tropical regions towards the poles (SMITH, 1962), the dependence on humidity, absorbed directly from the air, makes moist forests centers of epiphytic biodiversity (BENZING, 199; SCHÜTZ-GATTI, 2000; KERSTEN; SILVA, 2001). This may contribute to drier forest formations, as in the case of the sites studied, although at lower latitudes, being less diverse than wet forests at higher latitudes.

The quantitative assessment of the vascular epiphytes from all the sites (core and its replicates) in the Downstream Area is shown in Table 17. The downstream area of the Sorocaba/Médio Tiete basin has well-defined climatic characteristics, especially with regard to the distribution of moisture during the year - with colder, drier winters and wetter summers. It is possible to see the greater importance of species that are resistant to this period of water deficit, as is the case with some genera of the Polypodiaceae and Bromeliaceae families, especially the genera *Tillandsia* and *Pleopeltis,* which manage to overcome or resist periods of water deficit, and are precisely responsible for almost 44% of the epiphytic importance value (VIE) in the Downstream Area.

Table 17 - Vascular epiphytes of the Downstream Area of the Sorocaba/Médio Tietê Hydrographic Basin (Seasonal Semideciduous Forest/Cerrado Ecotone), classified according to the value of epiphytic importance - nr: absolute number of occurrences in the strata; far: absolute frequency in the strata; ni: absolute number of occurrences in the forophytic individuals; fai: absolute frequency in the forophytic individuals; vt (total value): sum of the abundance estimates; vie: epiphytic importance value; nota: average score obtained.

Species	nr	far	ni	fai	vt	vie
Pleopeltis pleopeltifolia	386	17.87	122	33.89	616	12.55
Tillandsia recurvata	376	17.41	125	34.72	607	12.37

Table 17- Continuation...

Species	nr	far	ni	fai	vt	vie
Microgramma squamulosa	319	14.77	97	26.94	573	11.67
Tillandsia tricholepis	298	13.80	99	27.50	500	10.19
Peperomia rotundifolia	158	7.31	79	21.94	345	7.03
Rhipsalis teres	153	7.08	63	17.50	306	6.23
Lepismium lumbricoides	130	6.02	57	15.83	254	5.17
Pleopeltis squalida	119	5.51	40	11.11	223	4.54
Microgramma tecta	117	5.42	42	11.67	184	3.75
Microgramma persicariifolia	87	4.03	33	9.17	155	3.16
Lepismium cruciforme	79	3.66	37	10.28	151	3.08
Epiphyllum phyllanthus	68	3.15	34	9.44	141	2.87
Philodendron bipinnatifidum	51	2.36	24	6.67	110	2.24
Pleopeltis hirsutissima	47	2.18	20	5.56	76	1.55
Ornithocephalus myrticola	36	1.67	16	4.44	66	1.34
Rhipsalis cereuscula	27	1.25	11	3.06	64	1.30
Tillandsia stricta	37	1.71	14	3.89	57	1.16
Vriesea bituminosa	26	1.20	14	3.89	50	1.02
Aechmea bromeliifolia	22	1.02	12	3.33	49	1.00
Tradescantia albiflora	27	1.25	13	3.61	45	0.92
Tillandsia funckiana	27	1.25	12	3.33	41	0.84
Peperomia trineura	18	0.83	11	3.06	34	0.69
Baptistonia lietzei	16	0.74	9	2.50	29	0.59
Acanthostachys strobilacea	14	0.65	8	2.22	28	0.57
Octomeria crassifolia	12	0.56	8	2.22	26	0.53
Rodriguezia decorates	12	0.56	4	1.11	22	0.45
Philodendron appendiculatum	11	0.51	6	1.67	21	0.43
Tillandsia usneoides	9	0.42	4	1.11	15	0.31
Bulbophyllum plumosum	7	0.32	4	1.11	13	0.26

Bulbophyllum chloroglossum	4	0.19	2	0.56	12	0.24
Serpocaulon latipes	9	0.42	4	1.11	12	0.24
Acianthera saundersiana	4	0.19	2	0.56	11	0.22
Peperomia tetraphylla	6	0.28	3	0.83	11	0.22
Octomeria gracilis	4	0.19	2	0.56	10	0.20
Rodriguezia sp.	6	0.28	3	0.83	9	0.18
Bulbophyllum epiphytum	3	0.14	3	0.83	7	0.14
Aechmea distichantha	2	0.09	1	0.28	5	0.10
Rhipsalis baccifera	4	0.19	2	0.56	5	0.10
Vittaria lineata	3	0.14	2	0.56	5	0.10
Campyloneurum sp.	2	0.09	2	0.56	4	0.08
Octomeria palmyrabellae	2	0.09	1	0.28	4	0.08
Epidendrum rigidum	2	0.09	1	0.28	3	0.06
Acianthera nemorosa	2	0.09	1	0.28	3	0.06
Peperomia glabella	2	0.09	1	0.28	3	0.06
Oeceoclades maculata	1	0.05	1	0.28	2	0.04
Polystachya foliosa	1	0.05	1	0.28	2	0.04

The species with the highest importance value was Pleopeltis *pleopeltifolia* (Polypodiaceae) with an epiphytic importance value (EVI) of 12.55 and an average score of 1.60; this species occurred in 33.9% of the forophytes and 17.9% of the strata. Tillandsia recurvata (Bromeliaceae), with a VIE of 12.37 and an average score of 1.61, occurring in around 35% of the forophytes and 17% of the strata, was the second most important species in the Downstream Area. *Microgramma squamulosa* (Polypodiaceae) had a VIE of 11.61 and an average score of 1.80, and was found in 26.9% of the forophytes and 14.7% of the strata; in addition, *Tillandsia tricholepis* had a VIE of 10.91 and an average score of 1.68, occurring in 27.5% of the forophytes and 13.8% of the strata. *Peperomia rotundifolia* (Piperaceae) had a VIE of 7.03, while *Rhipsalis teres* and *Lepismium Iumbricoides* (Cactaceae) had VIEs of 6.23 and 5.17, respectively. These seven species were responsible for more than 65% of the epiphytic importance value in the Ecotone between Semideciduous Seasonal Forest and Cerrado in the Downstream Area of the Sorocaba/Médio Tietê basin. The Polypodiaceae, Bromeliaceae and Cactaceae families are among the most frequently observed in Brazilian studies of Semideciduous Seasonal Forest (DISLICH; MANTOVANI, 1998; ROGALSKI; ZANIN, 2003; GIONGO; WAECHTER, 2004; BREIER, 2005; DETTKE et al. 2008; BATAGHIN et al., 2010) and Cerrado areas (BREIER, 2005; BATAGHIN et al., 2012b). The Polypodiaceae family was the most important in the area with a VIE of 37.54, followed by Bromeliaceae with a VIE of 27.54% and species from the Cactaceae family, responsible for almost 19% of the VIE of the Downstream Area. Resistance to water deficit and/or temperature variation may be responsible for the success of these families in this type of environment.

There was a significant difference between the shape of the vertical distribution of the abundances of vascular epiphytes that occurred in the Core Site and the Replica I Site (p < 0.05), but there was no significant difference (p > 0.05) between the vertical distribution of the Core Site in the Downstream Area and that recorded in the Replica II and Replica III Sites. The comparison between the same strata of the Core Site (UC) and the three Replica Sites revealed that the vertical distribution of the epiphytic community of the Core Site is different from that of the Replica I Site in all strata, except for the Inner Cup. However, the analysis revealed no difference between the vertical distribution occurring at the Core Site and that recorded at Replica Sites II and III, except for the Inner Cup of Replica Site III, which was significantly different from the Inner Cup of the Core Site in the Downstream Area.

With regard to epiphytic distribution in the strata, looking at all the sites in the Downstream Area, the base of the canopy stood out as the stratum with the greatest epiphytic abundance (Figure 18), with an abundance value (VA) of 1494, followed by the inner canopy with VA = 1154, the high stem with VA = 865 and the middle stem, outer canopy and low stem with abundance values of 554, 429 and 413, respectively.

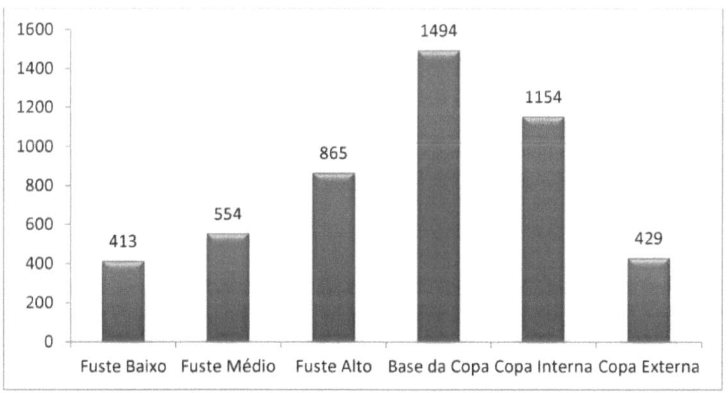

Figura 18:   Distribution of the abundance of epiphytic vascular species among the forophytic strata in the downstream area of the Sorocaba/Médio Tietê river basin.

The analysis of the vertical distribution of the epiphytic community in the Downstream Area, based on species abundance (Table 18), showed significant differences between the base of the canopy and three strata (low stem, medium stem and outer canopy) and between the inner canopy and two other strata (low stem and outer canopy). In terms of richness, the outer canopy showed a distinct epiphytic community from that occurring in all the other strata and was significantly different from the low canopy.

Table 18 - Analysis of the distribution of the richness and abundance of vascular epiphytes among the forophytic strata in the Downstream Area of the Sorocaba/Médio Tietê hydrographic basin.

^^^^^^Wealth Abundance^^^^	Beam Bass	Beam Medium	High stem	World Cup base	Internal Cup	Cup External
Low stem		0,1314	0,0824	0,003124	0,1991	2,89E-04
Medium stem	0,285		0,8196	0,1329	0,8237	5,99E-07
High stem	0,056	0,126		0,2023	0,6487	2,09E-07
World Cup base	0,004	0,009	0,067		0,08471	4,89E-10
Internal Cup	0,029	0,058	0,237	0,246		1,64E-06
External Cup	0,48	0,346	0,103	0,01	0,049	

Several authors have postulated that epiphytes show an irregular distribution along the forophytes, with the number of individuals varying vertically (STEEGE; CORNELISSEN, 1989; BROWN, 1990; WAECHTER, 1992). In the downstream area, this tendency was noticed, although the lack of difference between the base of the canopy and the inner canopy in terms of abundance and the differences between these two strata in relation to the others, can be attributed to the existence of greater availability of sites for the establishment of epiphytes (KERSTEN, 2006). In addition, the so-called "vertical evolution", in which epiphytes move to lower spaces in search of more light and conditions for acquiring water and nutrients (KIRA; YODA, 1989; BENZING, 1990) certainly has an influence on the abundance and richness of each zone. The distinct epiphytic community that occurs in the outer canopy corroborates this idea. The record of a small group of species adapted to this environment of great climatic variation (outer canopy), especially in terms of water acquisition, indicates the importance of climatic and microclimatic factors for the epiphytic community. The different communities that occur on the forophytes can also be seen in Figure 19, which shows the similarity (Jaccard) between the strata.

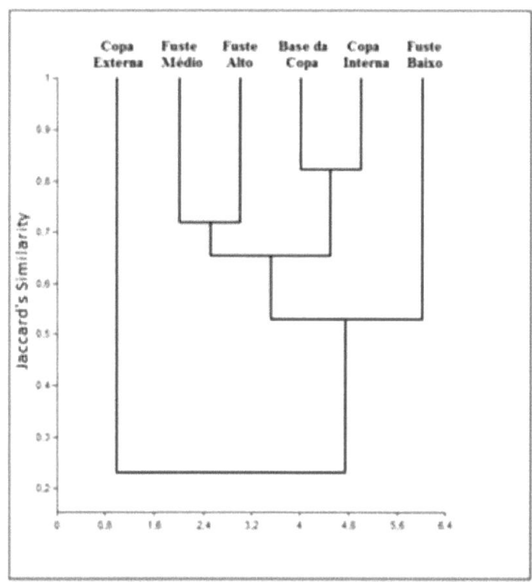

Figura 19: Dendrogram (UPGMA) of Jaccard's similarity between strata in the Semideciduous Seasonal Forest/Cerrado Ecotone in the downstream area of the Sorocaba/Médio Tietê basin.

The greater similarity between the base of the canopy and the inner canopy may be related to the presence of sites that tend to have a greater accumulation of suspended soil, which increases the retention of nutrients and moisture, favoring the establishment and development of epiphytes (BENZING, 1990; KROMER et al., 2007). However, not only in this result, but in all the sites in the downstream area, these two strata tended not to differ significantly from each other, indicating that there is a group of species adapted to the microclimatic conditions characteristic of this height of forophyte. In the same way, the epiphytic community recorded close to the ground and that which occurs in the outer canopy tends to be distinct from the other strata. It is possible to say that for the Downstream Area there are species better adapted to shaded areas with higher humidity (low canopy), an intermediate community (medium canopy up to the inner canopy) and species adapted to surviving in more extreme conditions (outer canopy). Although the species recorded in the outer canopy occur in the different strata of the forophytes depending on the lighting conditions of the forest fragment.

CENTRAL AREA

Floristic survey of vascular epiphytes in the Central Basin Area

Sorocaba/Médio Tietê River Basin

In the floristic survey of the Semideciduous Seasonal Forest present in the Central Area of the Sorocaba/Médio Tietê river basin, 64 species were recorded, belonging to 32 genera and nine families (Table 19). The Shannon index for the epiphytic vascular community was $H'= 2.872$, the equability $J = 0.686$ and the Margalef richness index (d) was 7.605.

Table 19 - List of vascular epiphyte species found in the phytosociological survey of the central area of the Sorocaba Mèdio Tietê river basin and their respective Ecological Categories (EC) - HLC: Characteristic Holoepiphyte; HLF: Facultative Holoepiphyte; HLA: Accidental Holoepiphyte; HMP: Primary Hemiepiphyte; HMS: Secondary Hemiepiphyte. Dispersal form (Disp.) - Zo: Zoochoric; An: Anemochoric. Reg: HUFSCar herbarium registration number (Im: Digital image).

Family	Species	EC	Disp.	Reg.
ARACEAE				
1	Anthurium comtum Schott	HLF	Zo	8486
2	Philodendron bipinnatifidum Schott	HMS	Zo	Im
3	Philodendron eximium Schott	HMP	Zo	8488
ASPLENIACEAE				
4	Asplenium pulchellum Raddi	HLA	An	8498
BROMELIACEAE				
5	Acanthostachys strobilacea (Schult. f.) Klotzsch	HLC	Zo	8431
6	Aechmea apocalyptica Reitz	HLF	Zo	8455
7	Aechmea bromeliifolia (Rudge) Baker	HLC	Zo	8432
8	Aechmea distichantha Lem.	HLF	Zo	8503
9	Aechmea nudicaulis (L.) Griseb.	HLF	Zo	8434
10	Billbergia amoena (Lodd.) Lindl.	HLC	Zo	8379
11	Billbergia distachya (Vell.) Mez	HLC	Zo	8445
12	Billbergia porteana Brongn. ex Beer	HLC	Zo	8457
13	Billbergia zebrina (Herb.) Lindl.	HLC	Zo	8433
14	Tillandsia funckiana Baker	HLC	An	8454
15	Tillandsia recurvata (L.) L.	HLC	An	8429
16	Tillandsia sp.	HLC	An	8425
17	Tillandsia stricta Sol. ex Sims	HLC	An	8428
18	Tillandsia tricholepis Baker	HLC	An	8430
19	*Tillandsia usneoides* (L.) L.	HLC	An	8464

Family Species EC Disp. Reg.

BROMELIACEAE

20	*Vriesea fenestralis* Linden & André	HLC	An	8444
21	*Vriesea friburgensis* Mez	HLF	An	8426
22	*Vriesea procera* (Mart. ex Schult. & Schult.f.) Wittm.	HLF	An	8508

CACTACEAE

23	*Cereus alacriportanus* Pfeiff.	HLF	Zo	Im
24	*Epiphyllum phyllanthus* (L.) Haw.	HLC	Zo	8474
25	*Lepismium cruciforme* (Vell.) Miq.	HLC	Zo	8388
26	*Lepismium lumbricoides* (Lem.) Barthlott	HLC	Zo	8386
27	*Lepismium warmingianum* (K. Schum.) Barthlott	HLC	Zo	8390
28	*Rhipsalis cereuscula* Haw.	HLC	Zo	8383
29	*Rhipsalis pilocarpa* Loefgr.	HLC	Zo	8385
30	*Rhipsalis floccosa* Salm-Dyck ex Pfeiff.	HLC	Zo	8380
31	*Rhipsalis teres* (Vell.) Steud.	HLC	Zo	8384
32	*Rhipsalis trigona* Pfeiff.	HLC	Zo	8381

COMMELINACEAE

33	*Tradescantia albiflora* Kunth	HLA	Zo	8391

ORCHIDACEAE

34	*Baptistonia lietzei* (Regel) Chiron & V.P.Castro	HLC	An	8407
35	*Brasiliorchis chrysantha* (Barb. Rodr.) R.B.Singer *et al.*	HLC	An	8397
36	*Cyclopogon multiflorus* Schltr.	HLA	An	8404
37	*Encyclia oncidioides* (Lindl.) Schltr.	HLC	An	8466
38	*Epidendrum rigidum* Jacq.	HLC	An	8453
39	*Lophiaris pumila* (Lindl.) Braem	HLC	An	8468
40	*Miltonia flavescens* (Lindl.) Lindl.	HLC	An	8412
41	*Oeceoclades maculata* (Lindl.) Lindl.	HLA	An	8400
42	*Polystachya estrellensis* Rchb.f	HLC	An	8467
43	*Polystachya foliosa* (Lindl.) Rchb.f.	HLC	An	8471
44	*Rodriguezia decora* Rchb. f.	HLC	An	8441
45	*Sophronitis cernua* Lindl.	HLC	An	Im

PIPERACEAE

46	*Peperomia trineuroides* Dahlst.	HLC	Zo	8450
47	*Peperomia glabella* (Sw.) A. Dietr.	HMP	Zo	8475
48	*Peperomia pereskiifolia* (Jacq.) Kunth	HMP	Zo	8448
49	*Peperomia tetraphylla* (G. Forst.) Hook. & Arn.	HLC	Zo	8447

POLYPODIACEAE

50	*Campyloneurum centrobrasilianum* Lellinger	HLC	An	8526
51	*Campyloneurum nitidum* (Kaulf.) C. Presl	HLC	An	8531

52	*Campyloneurum repens* (Aubl.) C. Presl	HLC	An	8533
53	*Microgramma persicariifolia* (Schrad.) C. Presl	HLC	An	8520
54	*Microgramma tecta* (Kaulf.) Alston	HLC	An	8524

Family Species EC Disp. Reg.				

POLYPODIACEAE

55	*Microgramma squamulosa* (Kaulf.) de la Sota	HLC	An	8534
56	*Microgramma vacciniifolia* (Langsd. & Fisch.) Copel.	HLC	An	8522
57	*Pecluma filicula* (Kaulf.) M.G. Price	HLC	An	8527
58	*Pleopeltis astrolepis* (Liebm.) E. Fourn.	HLC	An	8519
59	*Pleopeltis hirsutissima* (Raddi) de la Sota	HLC	An	8521
60	*Pleopeltis pleopeltifolia* (Raddi) Alston	HLC	An	8525
61	*Pleopeltis squalida* (Vell.) de la Sota	HLC	An	8436
62	*Serpocaulon latipes* (Langsd. & Fisch.) A.R. Sm.	HLF	An	8423

PTERIDACEAE

63	*Polytaenium cajenense* (Desv.) Benedict	HLC	An	8439
64	*Vittaria lineata* (L.) Sm.	HLC	An	8454

The richness of epiphytic species found in the area can be considered similar to that observed in the surveys in Semideciduous Seasonal Forest carried out by Rogalski and Zanin (2003) who found 70 species, by Giongo and Waechter (2004) who sampled 57 species; by Cervi and Borgo (2007) who found 56 species; by Alves and Kolbek (2009) with 61 species; by Menini-Neto et al. (2009) with 59 species (Descoberto-MG); by Geraldino et al. (2010) with 61 species and by Bonnet et al. (2011) with 60 species. However, it can be considered higher than that observed in the surveys carried out by Aguiar et al. (1981), who sampled 17 species; by Dislich and Mantovani (1998), 34 species; by Borgo et al. (2002), with 32 species; by Breier (2005), 25 species; by Dettke et al. (2008), 29 species; by Menini-Neto et al. (2009), with 41 species (Barroso-MG) and by Bataghin *et al.* (2010) with 21 species.

When compared to wetter forest formations, the specific richness of the area can be considered low, since in Dense Ombrophilous Forest several authors report a higher number of epiphyte species, e.g. Blum (2010) - 277 species; Petean (2009) - 159 species; Kersten (2006) - 349 species; Breier (2005) - 161 species; Fontoura et al. (1997) - 293 species; Hertel (1950) - 101 species; Schütz-Gatti (2000) - 175 species and Petean (2003) - 97 species. These much higher numbers found in Dense Ombrophilous Forest reinforce the idea that epiphytes are dependent on atmospheric humidity (GENTRY; DODSON, 1987a), since the acquisition and storage of water are the most important factors for epiphytic growth (ZOTS; HIETZ 2001).

In the survey of the Semideciduous Seasonal Forest in the central area of the Sorocaba/Médio Tietê river basin, the epiphytic families with the highest species richness were: Bromeliaceae (18 species), Polypodiaceae (13 species), Orchidaceae (12 species) and Cactaceae (eight species). The Aspleniaceae and Commelinaceae families had only one species each. The distribution of epiphytic species in the ecological categories (Figure 20), according to the relationship with the forophyte proposed by Benzing (1990), showed the predominance of characteristic holoepiphytes with 48 species (73%); followed by facultative holoepiphytes with eight species (13%); accidental holoepiphytes with five species (8%), primary hemiepiphytes with four species (6%) and secondary hemiepiphytes with only one species (2%). The predominance of characteristic holoepiphytes has been observed in Semideciduous Seasonal Forest in several studies (PINTO et al., 1995; DISLICH; MANTOVANI, 1998; ROGALSKI; ZANIN, 2003; CERVI; BORGO, 2007, DETTKE et al, 2008), and in other forest formations, such as mixed ombrophilous forests (DITTRICH et al., 1999) and restinga areas (WAECHTER, 1992, KERSTEN; SILVA, 2001).

The classification of vascular epiphytes according to dispersal syndrome in the central area highlighted 56% of the species as anemochoric and 44% as zoochoric. Anemochory has predominated as a dispersal syndrome among epiphytic species in several studies (BENZING, 1987; BREIER, 2005; DETTKE et al., 2008, MENINI- NETO et al. 2009; GERALDINO et al. 2010). However, (BENZING, 1987) points out that around two thirds of vascular epiphyte species show anemochoric dispersal, a different pattern to that observed in this area of the basin. One possible explanation for the reduction in the number of anemochoric species is the occurrence of forest fragments that are totally dispersed in the anthropogenic matrix and of low structural complexity. The distance from other forest fragments represents a barrier to anemochoric dispersal and consequent colonization of new areas, and low structural complexity does not allow the development of anemochoric species that are more demanding in microclimatic terms, as is the case with orchids. This low structural complexity contributes to the development of species that are more adapted to conditions of great climatic variability, such as the Cactaceae, which have zoochoric dispersal. This lower proportion of anemochoric epiphytic species was also observed in the Downstream Area, in isolated forest fragments with low structural complexity.

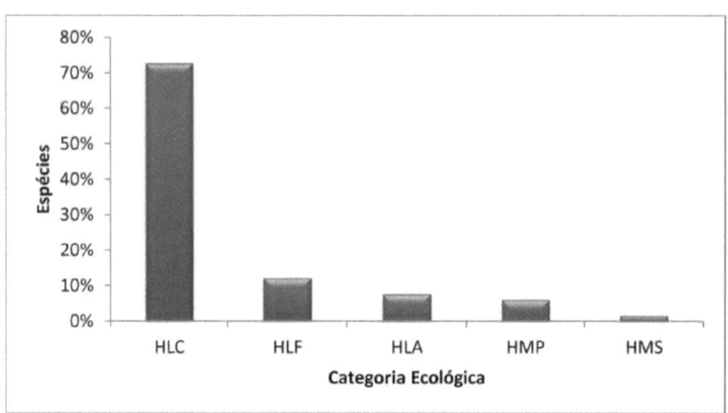

Figura 20: Distribution of the epiphytic vascular species of the Seasonal Semideciduous Forest in the Central Area of the Sorocaba/Médio Tietê hydrographic basin in the ecological categories proposed by Benzing (1990) - HLC: characteristic holoepiphytes; HLF: facultative holoepiphytes; HLA: accidental holoepiphytes; HMP: primary hemiepiphytes; HMS: secondary hemiepiphytes.

The Orchidaceae, Bromeliaceae and Polypodiaceae families, which are the richest in epiphytes worldwide (MADISON, 1977; KRESS, 1986; GENTRY; DODSON, 1987b; BENZING, 1990) are responsible for 65% (43sp.) of the species in the floristic survey, a percentage very similar to that found by Kersten (2006) in the Alto Iguaçu Basin, Paranà. Although the Cactaceae family accounts for around 0.5% of the world's epiphytic species (MADISON, 1977; BENZING, 1990) and 3% of Brazilian epiphytes (KERSTEN, 2006), it had 10 species (15%) in the study area. The great resistance of the Cactaceae to water stress due to their "relatively accentuated xeromorphism" (COUTINHO, 1962), an adaptation that probably offers benefits in relation to periods of water deficit, may be responsible for the high representativeness of this family.

Another aspect worth mentioning regarding the families found in the sites studied is the Orchidaceae family. Although it is the richest family worldwide (MADISON, 1977; KRESS, 1986; BENZING, 1990), in the Neotropics (GENTRY; DODSON, 1987a) and in Brazil (KERSTEN, 2006), it had 10 species, a smaller number than those found by Rogalski and Zanin (2003) and Cervi and Borgo (2007), but very similar to the results obtained by Dislich and Mantovani (1998), Borgo et al. (2002) and higher than those of Breier (2005), Dettke et al. (2008), Bataghin et al. 2010. Stancato et al. (2002) suggest that high light intensity can reduce the growth and development of orchids. The climatic variability characteristic of the study area, which usually has winters with prolonged droughts, as well as the deciduousness of the forest itself, which makes the canopy less thick and allows more light to pass through, has an influence on these numbers. In addition, Ditt (2002) highlights anthropogenic interference, microclimatic changes resulting from environmental modification and the possible illegal collection of specimens (especially those of commercial interest - ornamental), as factors that can lead to a reduction in the number of individuals, contributing to a decrease in biological diversity and environmental degradation.

Distribution of vascular epiphytes in the sampling sites in the Central Area

Analysis of the vascular epiphytes of the Core Site of the Semideciduous Seasonal Forest in the Central Area of the Sorocaba/Médio Tietê river basin

The survey of vascular epiphytes carried out at the Core Site in the central area took place in the conservation unit known as the Ipanema National Forest, located between UTM coordinates 232.187 and 7.406.164 in zone 23 S, which covers 5,179.93 hectares, most of which is in the southern part of the municipality of Iperó-SP (FAVERO et al. 2004). The Ipanema National Forest was created in May 1992 (IBAMA 2003) and is an extensive area of Semideciduous Seasonal Forest, the remnants of a totally fragmented vegetation with a long history of disturbance. The area where the Ipanema National Forest is located was home to the first Brazilian steel mill (Figure 21), as well as apatite extraction for the manufacture of superphosphate, limestone extraction for the production of cement, seed and agricultural machinery trials and even today a research center for the development of nuclear reactors for submarines (IBAMA 2003).

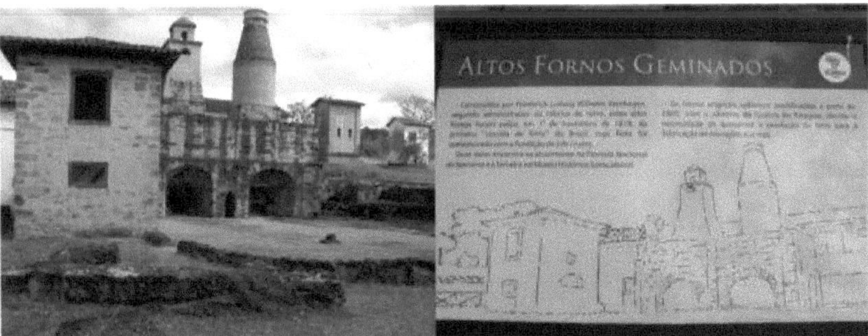

Figura 21: Old iron smelting furnaces present in the Conservation Unit, which used coal produced from the forest of the current Fiona de Ipanema.

In the floristic survey of Sitio Core, 21 species were recorded, belonging to 13 genera and six families (Table 20). The Shannon index of the site was H' = 2.335, the equability (J) was 0.767 and the Margalef richness (d) was 2.956.

Table 20 - List of vascular epiphyte species found in the Core Site of the Central Area (Ipanema National Forest) of the Sorocaba/Médio Tietê hydrographic basin and their respective Ecological Categories (EC) - HLC: Characteristic holoepiphyte; HLF: Facultative holoepiphyte; HLA: Accidental holoepiphyte; HMP: Primary hemiepiphyte; HMS: Secondary hemiepiphyte. Dispersal form (Disp.) - Zo: Zoochoric; An: Anemochoric. Reg: HUFSCar herbarium registration number (Im: Digital image).

Family	Species	EC	Disp.	Reg.
ARACEAE				
1	Anthurium comtum Schott	HLF	Zo	8486
2	Philodendron bipinnatifidum Schott	HMP	Zo	Im
BROMELIACEAE				
3	Aechmea apocalyptica Reitz	HLF	Zo	8455
4	Aechmea bromeliifolia (Rudge) Baker	HLC	Zo	8432
5	Aechmea distichantha Lem.	HLF	Zo	8445
6	Tillandsia recurvata (L.) L.	HLC	An	8429
7	Tillandsia stricta Sol. ex Sims	HLC	An	8428
8	Tillandsia tricholepis Baker	HLC	An	8430
CACTACEAE				
9	Cereus alacriportanus Pfeiff.	HLF	Zo	Im
10	Epiphyllum phyllanthus (L.) Haw.	HLC	Zo	8474
11	Lepismium cruciforme (Vell.) Miq.	HLC	Zo	8388
12	Lepismium lumbricoides (Lem.) Barthlott	HLC	Zo	8386
13	Lepismium warmingianum (K. Schum.) Barthlott	HLC	Zo	8390
14	Rhipsalis cereuscula Haw.	HLC	Zo	8383
15	*Rhipsalis teres* (Vell.) Steud.	HLC	Zo	8384

Family Species EC Disp. Reg.

Family	Species	EC	Disp.	Reg.
COMMELINACEAE				
16	*Tradescantia albiflora* Kunth	HLA	Zo	8391
ORCHIDACEAE				
17	*Epidendrum rigidum* Jacq.	HLC	An	8453
18	*Oeceoclades maculata* (Lindl.) Lindl.	HLA	An	8400
POLYPODIACEAE				
19	*Microgramma squamulosa* (Kaulf.) de la Sota	HLC	An	8534
20	*Pleopeltis pleopeltifolia* (Raddi) Alston	HLC	An	8525
21	*Pleopeltis squalida* (Vell.) de la Sota	HLC	An	8436

Given the extent of the Sitio Core area and the fact that it is a Conservation Unit, the richness of epiphytic species is low, especially if we remember the studies carried out in

Semideciduous Seasonal Forest by Rogalski and Zanin (2003) who found 70 species, by Giongo and Waechter (2004) who sampled 57 species and by Cervi and Borgo (2007) who found 56 species. Despite this, it can be considered similar, in this same type of forest, to that observed by Aguiar et al. (1981), who sampled 17 species; by Dislich and Mantovani (1998), 34 species; by Borgo *et al.* (2002), with 32 species; by Breier (2005), 25 species and by Dettke et al. (2008), 29 species. This low richness of Sitio Core is certainly influenced by the long period of anthropogenic interference suffered by the forest (IBAMA 2003). The commercial exploitation of trees (many of which were used as fuel in the steel mill's furnaces) and deforestation for mineral extraction, as well as seed and machinery tests (including aerial pesticide dispersal tests), undoubtedly had an influence on the vascular epiphytic community at this site, since the loss of epiphytic diversity as a result of human interference in environments has been reported by several authors (ENGWALD et al., 2000; BARTHLOTT et al., 2001; WOLF, 2005; BATAGHIN et al., 2008; DETTKE et al., 2008).

In the Core Site, Cactaceae and Bromeliaceae were the richest families, with seven and six species, respectively. The Polypodiaceae family (the richest family in the core area) had three species and the Commelinaceae family only one species. Characteristic holoepiphytes were dominant with 67% of the species, followed by facultative holoepiphytes (19%) and accidental holoepiphytes (10%), while primary hemiepiphytes accounted for 4% of the species.

Worldwide, around two thirds of vascular epiphyte species show anemochoric dispersal (BENZING, 1987), and this inversely proportional pattern was observed at this site in the central area of the basin. The classification of species according to dispersal syndrome recorded 65% of them as zoochoric and 35% as anemochoric. Although anemochory has predominated as a dispersal syndrome among epiphytic species (BENZING, 1987; BREIER, 2005; DETTKE et al., 2008, MENINI-NETO et al., 2009; GERALDINO et al., 2010), there is a great possibility that the reduction in the number of anemochoric species is associated with the occurrence of forest fragments that are totally dispersed in the anthropogenic matrix and also with the low structural complexity of these fragments. The distance from other forest fragments represents a barrier to anemochoric dispersal and consequent colonization of new areas, and the low structural complexity does not allow the development of anemochoric species that are more demanding in microclimatic terms, as is the case with orchids. This low structural complexity contributes to the development of species that are more adapted to conditions of great climatic variability, such as the Cactaceae, which have zoochoric dispersal and were the richest family in this site. A reduction in the proportion of anemochoric epiphytic species was also observed in the downstream area in isolated forest fragments of low structural complexity, although this was the first case of an inversion of the proportions of dispersal syndromes, with zoochory predominating.

The quantitative analysis recorded 21 species and highlighted two Bromeliaceae as the most important species for the Core Site (Table 21): *Tillandsia tricholepis* (Figure 22) and *Tillandsia recurvata* were responsible for almost 40% of the epiphytic importance value (VIE) recorded at this site. The first had a VIE of 20.98 and an average score of 1.54, being recorded in more than 64% of the forophytes and almost 36% of the strata. The second species had a VIE of 18.28 and an average score of 1.57, as well as occurring in 57.8% of the forophytes and 30.2% of the strata. This contributed to the Bromeliaceae family, with 47.7% of the VIE, being highlighted as the most important in Sitio Core. These two species are poikilohydric, showing great resistance to lack of humidity and extreme conditions of temperature and light, and are characteristic of secondary vegetation and/or vegetation under human influence (DETTKE et al., 2008) and also in Cerrado areas (BATAGHIN et al. 2012b). The great success of these species is indicative of the vegetation that occurs on this site, which is more permeable to solar radiation and has more restrictive conditions in terms of water availability.

Table 21 - Vascular epiphytes from the Core Site of the Central Area in the Sorocaba Médio Tietê hydrographic basin, classified according to the value of epiphytic importance - nr: absolute number of occurrences in the strata; far: absolute frequency in the strata; ni: absolute number of occurrences in the forophytic individuals; fai: absolute frequency in the forophytic individuals; vt (total value): sum of the abundance estimates; vie: value of epiphytic importance; grade: average grade obtained.

Species	nr	far	ni	fai	vt	vie	note
Tillandsia tricholepis	192	35,6	58	64,4	295	20,98	1,54
Tillandsia recurvata	163	30,2	52	57,8	257	18,28	1,58
Pleopeltis pleopeltifolia	113	20,9	44	48,9	191	13,58	1,69
Pleopeltis squalida	75	13,9	22	24,4	139	9,89	1,85
Tillandsia stricta	70	13,0	30	33,3	100	7,11	1,43
Rhipsalis cereuscula	49	9,1	25	27,8	92	6,54	1,88
Microgramma squamulosa	59	10,9	21	23,3	91	6,47	1,54
Epiphyllum phyllanthus	52	9,6	24	26,7	79	5,62	1,52
Lepismium warmingianum	19	3,5	8	8,9	34	2,42	1,79
Rhipsalis teres	11	2,0	8	8,9	26	1,85	2,36
Lepismium cruciforme	15	2,8	8	8,9	24	1,71	1,60

Cereus alacriportanus	8	1,5	4	4,4	17	1,21	2,13
Aechmea distichantha	11	2,0	5	5,6	15	1,07	1,36
Epidendrum rigidum	7	1,3	4	4,4	10	0,71	1,43
Lepismium lumbricoides	7	1,3	4	4,4	10	0,71	1,43
Philodendron bipinnatifidum	6	1,1	4	4,4	8	0,57	1,33
Anthurium comtum	4	0,7	2	2,2	7	0,50	1,75
Oeceoclades maculata	3	0,6	2	2,2	5	0,36	1,67
Aechmea apocalyptica	1	0,2	1	1,1	2	0,14	2,00
Aechmea bromeliifolia	1	0,2	1	1,1	2	0,14	2,00
Tradescantia albiflora	2	0,4	1	1,1	2	0,14	1,00

The species of the Polypodiaceae family, especially *Pleopeltis pleopeltifolia,* with a VIE = 13.58 and an average score of 1.69 and *Pleopeltis squalida with a VIE* = 9.89 and an average score of 1.85, helped make this family the second most important for this site, with a total VIE of 29.94. Although the Cactaceae family had the highest number of epiphytic species, it only had the third highest importance value - 20.06. Dislich and Mantovani (1998) highlight the neotropical endemism of Bromeliaceae and Cactaceae, which may be one of the reasons for the numbers presented here; in addition, according to Scheinvar (1985), the genus *Rhipsalis (*Cactaceae), for example, has its dispersal center in the south and southeast of Brazil, which favors the occurrence of species from this family in the area studied.

Figura 22:   Tillandsia tricholepis Baker (Bromeliaceae), the most abundant species in the Core Site of the Central Area, is highly resistant to periods of water deficit. A - habit; B - flower.

The low number of species of vascular epiphytes is a characteristic of forests that suffer from anthropogenic interference (BATAGHIN et al., 2008), corroborating this idea, Barthlott et al. (2001) observed a reduction in the number of species due to an increase in the degree of human interference and the successional stage of the forest as a result of such interference. In addition, Dettke et al. (2008) observed a concentration of species in the Bromeliaceae,

Cactaceae and Polypodiaceae families (which are resistant to environments with more light and less humidity) in an altered remnant of Semideciduous Seasonal Forest. In the Core Site in the central area, these three families are responsible for the largest number of species and the highest values of epiphytic importance, confirming the probable environmental changes in the site.

The distribution of epiphytes in the forophyte strata at Sitio Core (Figure 23) showed that the base of the canopy, with an abundance value (VA) of 552, was the stratum with the highest epiphyte abundance; the second most abundant stratum was the inner canopy with a VA of 346, followed by the high stem with a VA of 223, the middle stem with a VA of 142, the outer canopy with a VA of 77 and the low stem with a VA of 56.

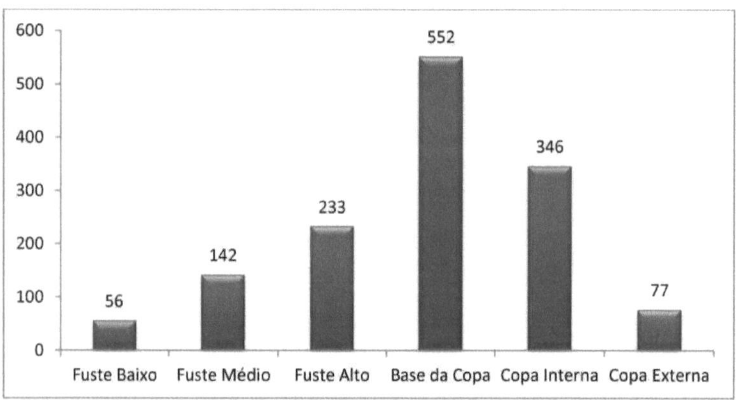

Figura 23:   Distribution of the abundance of epiphytic vascular species among the forophytic strata in the Core Site of the central area in the Sorocaba/Médio Tietê hydrographic basin.

The statistical analysis applied to the vertical distribution of vascular epiphytes in the forophytic strata revealed a significantly different abundance between the extreme strata (low and medium stem; outer canopy) compared to the intermediate strata (Table 22). Similarly, the richness of the low stem and outer canopy (ends of the forophyte) were significantly different in relation to the intermediate parts of the host trees. This occurred at this site because, although the species recorded in the low stem and outer canopy also occur in the other strata, the environmental factors (microclimatic or even support factors) mean that the epiphytic communities that develop in these two regions are made up of a few species, generally those with greater plasticity.

This type of richness distribution, with the same species occurring in the low stem and outer canopy, can be characterized as typical of impacted forests or forests in an intermediate successional stage. Further evidence of this is the small number of individuals in the low stem, since in more degraded forests this stratum tends to be less abundant. On the other hand, in the intermediate regions of the forophytes, the presence of branches - which favor epiphytism by providing better sites for installation and survival in microclimatic terms - allows for the development of a more diverse vascular epiphytic community.

Table 22: Analysis of the distribution of the richness and abundance of vascular epiphytes among the forophytic strata in the Core Site of the Central Area in the Sorocaba/Médio Tietê hydrographic basin.

^^^-^^^Rich Abundance'~~~~~^	Beam Bass	Beam Medium	Beam High	World base	Cup	Cup Internal	Cup External
Low stem		**0,01012**	**0,01012**	**0,003154**		0,124	0,2147
Medium stem	0,034		0,9834	0,6539		0,2727	**2,31E-04**
High stem	**0,006**	0,119		0,6539		0,2727	**2,31E-04**
World Cup base	**4,02E-04**	**0,003**	**0,02**			0,1249	**5,75E-05**
Internal Cup	**0,009**	0,051	0,198	0,126			**0,00634**
External Cup	0,305	0,118	**0,02**	**7,70E-04**	**0,015**		

The pattern of vertical distribution found at Sitio Core da Are Central reflects the adaptation of vascular epiphytes to the great microclimatic variations that exist, in which they exchange spaces in search of better conditions for acquiring water and nutrients (KIRA; YODA, 1989; BENZING, 1990), although in certain types of forest, such as Semideciduous, this limits the occurrence of environmentally more sensitive species. For this site, the importance of places that can serve as suspended soil deposits and that contribute to the maintenance and acquisition of moisture by the epiphytes is also highlighted (KROMER et al., 2007), since this tends to be the main regulating factor of this community in the area.

Analysis of the vascular epiphytes of Sitio Rèplica I in the Central Area of the Sorocaba/Médio Tietê basin

The survey of vascular epiphytes at Sitio Rèplica I in Semideciduous Seasonal Forest in the central area of the watershed was carried out on a private rural property in the municipality of Boituva - SP. The area of the forest remnant is approximately 53 ha, located between UTM coordinates 234.690 and 7.413.837 in zone 23 South, and is characterized by being a Semideciduous Seasonal Forest and one of the largest forest fragments in the municipality. The steep slope of the forest remnant, which lies on the banks of the Sorocaba River, was fundamental to maintaining the vegetation cover. Although the relief of the area is very steep, there were strong indications of human presence (interference), such as the existence of trails and waiting (jirau) possibly used for hunting and fishing (Figure 24). The surrounding areas are characterized by anthropogenic agricultural use, with sugar cane and orange cultivation, as well as cattle and horse breeding.

Figura 24:   Interior of Sitio Rèplica I in the Central Area: A - jirau trap; B - trail used to access the Sorocaba River and slope of the area.

The floristic analysis of the area revealed the presence of 20 species, belonging to 12 genera and five families (Table 23). The Shannon index for the area was H' = 2.307, the equability (J) was 0.770 and the Margalef richness (d) was 2.849.

Table 23 - List of vascular epiphyte species found at Sitio Rèplica I in the central area (Fazenda Sitio Grande) of the Sorocaba Mèdio Tietê hydrographic basin and their respective Ecological Categories (EC) - HLC: Holoepiphyte characteristic; HLF: Holoepiphyte facultative; HLA: Holoepiphyte accidental; HMP: Hemiepiphyte primary; HMS: Hemiepiphyte secondary. Dispersal form (Disp.) - Zo: Zoochoric; An: Anemochoric. Reg: HUFSCar herbarium registration number (Im: Digital image).

N	Family	Species	EC	Disp.	Reg.
1	Araceae	Philodendron bipinnatifidum Schott	HMP	Zo	Im
2	Aspleniaceae	Asplenium pulchellum Raddi	HLA	An	8498
3	Bromeliaceae	Aechmea bromeliifolia (Rudge) Baker	HLC	Zo	8432
4	Bromeliaceae	Billbergia porteana Brongn. ex Beer	HLC	Zo	8457
5	Bromeliaceae	Billbergia distachya (Vell.) Mez	HLC	Zo	8445
6	Bromeliaceae	Tillandsia funckiana Baker	HLC	An	8458
7	Bromeliaceae	Tillandsia recurvata (L.) L.	HLC	An	8429
8	Bromeliaceae	Tillandsia tricholepis Baker	HLC	An	8430
9	Bromeliaceae	Vriesea fenestralis Linden & André	HLC	An	8444
10	Cactaceae	Epiphyllum phyllanthus (L.) Haw.	HLC	Zo	8474
11	Cactaceae	Lepismium cruciforme (Vell.) Miq.	HLC	Zo	8388

12	Cactaceae	Lepismium lumbricoides (Lem.) Barthlott	HLC	Zo	8386
13	Cactaceae	Rhipsalis cereuscula Haw.	HLC	Zo	8383
14	Cactaceae	Rhipsalis pilocarpa Loefgr.	HLC	Zo	8385
15	Polypodiaceae	Campyloneurum repens (Aubl.) C. Presl	HLC	An	8533
16	Polypodiaceae	Micrbogramma persicariifolia (Schrad.) C. Presl	HLC	An	8520

Keep going...

Table 23 - Continuation...

N	Family	Species	EC	Disp.	Reg.
17	Polypodiaceae	*Microgramma tecta* (Kaulf.) Alston	HLC	An	8524
18	Polypodiaceae	*Microgramma vacciniifolia* (Langsd. & Fisch.) Copel.	HLC	An	8522
19	Polypodiaceae	*Pleopeltis pleopeltifolia* (Raddi) Alston	HLC	An	8525
20	Polypodiaceae	*Pleopeltis squalida* (Vell.) de la Sota	HLC	An	8436

The richness of Sitio Rèplica I, considering that it is a small forest remnant in a predominantly anthropogenic landscape, can be compared to that observed in the works by Aguiar et al. (1981), who sampled 17 species; by Dislich and Mantovani (1998), 34 species; by Borgo et al. (2002), with 32 species; by Breier (2005), 25 species and by Dettke et al. (2008), 29 species. And above all, it can be compared to that found in the Core Site of this study, a Conservation Unit of over 5,000 ha. The absence of species from the Orchidaceae family, even though there is a humid microclimate on the site, can be considered an indication of the illegal collection of species. This type of anthropic action is highlighted by Ditt (2002), who presents the collection of species used for ornamentation, e.g. Orchidaceae and Bromeliaceae, as one of the factors leading to a reduction in the number of individuals and contributing to a decrease in biological diversity and environmental degradation.

The Bromeliaceae family was the richest at Sitio Rèplica I, with seven epiphytic species, followed by Polypodiaceae with six species and Cactaceae with five species. The Araceae and Aspleniaceae families had one epiphytic species each. In this site, characteristic holoepiphytes were the predominant form, accounting for 85% of the species in the area, while facultative holoepiphytes, accidental holoepiphytes and secondary hemiepiphytes accounted for 5% of the species each. The classification of vascular epiphytes according to dispersal syndrome recorded 55% anemochorous species and 45% zoochorous. Although anemochory is predominant in several studies (BREIER, 2005; DETTKE et al., 2008; MENINI-NETO et al., 2009; GERALDINO et al.., 2010) and has been recorded in the majority of species at this site, the reduction in the expected proportion, which is 2/3 anemochorous epiphytes (BENZING, 1987), may be a consequence of forest fragmentation and other anthropogenic factors that reduce the structural complexity of these forest areas, which contributes, for example, to the absence of orchids, which are the most numerous anemochorous group.

All 20 species found in the floristic survey were present in the quantitative study of Sitio Rèplica I (Table 24). *Microgramma Vacciniifolia* (Polypodiaceae) was the most important species on the site with an epiphytic importance value (VIE) of 20.71 and an average score of 1.83, being present in 32.2% of the forophytes and 16.5% of the strata. A Cactaceae, *Lepismium cruciforme* (Figure 25), was the second most important species with a VIE of 19.06 and an average score of 2.03. It was sampled in 26.7% of the forophytes and 13.7% of the forophytic strata. *Pleopeltis pleopeltifolia* (Polypodiaceae) with a VIE of 15.63 and an average score of 1.73, as well as being present in 27.8% of the forophytes and 13.1% of the strata, was the third most important species at Sitio Rèplica I in the central area.

Table 24 - Vascular epiphytes from Sitio Rèplica I in the central area of the Sorocaba Mèdio Tietê hydrographic basin, classified according to epiphytic importance value - nr: absolute number of occurrences in the strata; far: absolute frequency in the strata; ni: absolute number of occurrences in forophytic individuals; fai: absolute frequency in forophytic individuals; vt (total value): sum of abundance estimates; vie: epiphytic importance value; nota: average score obtained.

Species	nr	far	ni	fai	vt	vie	note
Microgramma vacciniifolia	89	16,5	29	32,2	163	20,71	1,83
Lepismium cruciforme	74	13,7	24	26,7	150	19,06	2,03
Pleopeltis pleopeltifolia	71	13,1	25	27,8	123	15,63	1,73
Aechmea bromeliifolia	52	9,6	26	28,9	98	12,45	1,88
Tillandsia recurvata	34	6,3	15	16,7	66	8,39	1,94
Vriesea fenestralis	28	5,2	10	11,1	48	6,10	1,71
Epiphyllum phyllanthus	12	2,2	6	6,7	22	2,80	1,83
Tillandsia funckiana	15	2,8	7	7,8	22	2,80	1,47
Billbergia porteana	11	2,0	8	8,9	19	2,41	1,73
Microgramma persicariifolia	9	1,7	2	2,2	19	2,41	2,11
Tillandsia tricholepis	8	1,5	5	5,6	15	1,91	1,88
Rhipsalis cereuscula	7	1,3	3	3,3	14	1,78	2,00
Lepismium lumbricoides	5	0,9	3	3,3	11	1,40	2,20
Pleopeltis squalida	2	0,4	1	1,1	5	0,64	2,50
Microgramma tecta	2	0,4	1	1,1	3	0,38	1,50
Asplenium pulchellum	2	0,4	1	1,1	3	0,38	1,50
Rhipsalis pilocarpa	2	0,4	2	2,2	2	0,25	1,00
Philodendron bipinnatifidum	2	0,4	1	1,1	2	0,25	1,00
Campyloneurum repens	1	0,2	1	1,1	1	0,13	1,00
Billbergia distachya	1	0,2	1	1,1	1	0,13	1,00

The Bromeliaceae *Aechmea bromeliifolia* with a VIE of 12.45 and an average score of 1.88, *Tillandsia recurvata* with a VIE of 8.39 and an average score of 1.94 and Vriesea fenestralis with a VIE of 6.10 and a score of 1.71 are also noteworthy. These species make the Bromeliaceae family the second most important in the site, accounting for 34.18% of the epiphytic importance value. The Polypodiaceae family, the most important, was responsible for 39.9% of the VIE even though it was not the richest in species.

Figura 25: Habitat of *Lepismium cruciforme* (Vell.) Miq. (Cactaceae), the species with the second highest epiphytic importance value.

The same trend of low numbers of species at Sitio Rèplica I was observed by Barthlott et al. (2001), Bataghin et al. (2008) and also at Sitio Core in the Central Area, which is characteristic of areas with anthropogenic interference. Added to this is the concentration of species in the Bromeliaceae, Polypodiaceae and Cactaceae families, reported by Dettke et al. (2008) as predominant in an altered remnant of Semideciduous Seasonal Forest.

Analysis of the forophytic strata revealed that the base of the canopy (Figure 26) was the most abundant with an abundance value (AV) of 274, the high stem was the second most abundant stratum with an AV of 180, followed by the inner crown with an AV of 177. The medium stem, outer crown and low stem had abundance values of 79, 54 and 23, respectively.

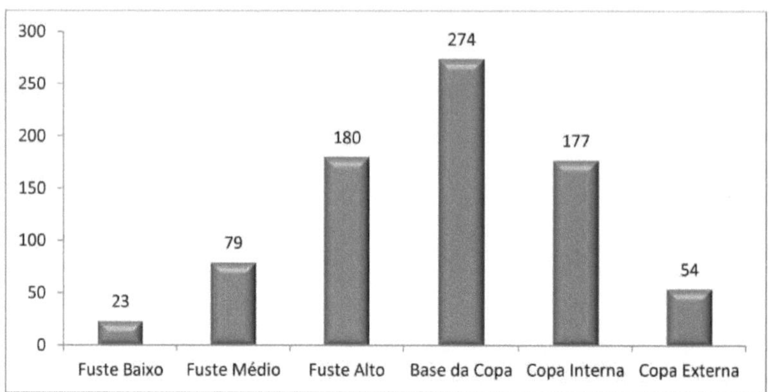

Figura 26:   Distribution of the abundance of epiphytic vascular species among the forophytic strata at Sitio Rèplica I in the central area of the Sorocaba/Médio Tietê hydrographic basin.

The distribution of vascular epiphytes in the forophytic strata based on abundance revealed that only the low stem does not show significant differences from the outer canopy. As can be seen in Figure 26, these strata have a small number of epiphytic individuals. Especially in the low canopy, the reduction in the number of individuals indicates a limitation to the development of the epiphytic community. In terms of richness, the outer canopy showed a different community from three other strata and the low stem differed from two strata (Table 25). This greater difference occurs because the intermediate region of the forophytes (high stem - inner crown) presents (in dry, less conserved forests or in intermediate successional stages) better conditions for the occurrence of epiphytic species that are more sensitive to large microclimatic variations (BATAGHIN et al. 2012a). In the low stem and outer canopy, almost half of the species recorded occur in all forophyte strata, but the more limited light and water conditions in these strata eliminate a large part of the total number of species recorded in the forest fragment.

Table 25 - Analysis of the distribution of the richness and abundance of vascular epiphytes among the forophytic strata at Sitio Réplica I in the Central Area of the Sorocaba/Médio Tietê hydrographic basin.

^^^-^^^Rich / Abundance'~~~^	Beam Bass	Beam Medium	Beam High	World base	Cup Internal	Cup External
Low stem		0,3515	**0,0047**	**0,0015**	0,0632	0,7515
Medium stem	**0,033**		0,0514	**0,0205**	0,3515	0,2081
High stem	**0,004**	0,056		0,6963	0,3023	**0,0018**
World Cup base	**0,003**	**0,018**	0,18		0,1544	**0,0005**
Internal Cup	**0,008**	0,074	0,485	0,178		**0,0298**

| External Cup | 0,184 | 0,282 | **0,029** | **0,01** | **0,04** |

There is a tendency for the abundance and richness of the epiphytes at Sitio Réplica I to be concentrated in the intermediate strata, especially at the base of the canopy, where there is an intermediate climatic gradient between water stress and access to light, as well as the presence of branch insertion points which favor the accumulation of suspended soil, contributing to the establishment of the species and also helping to retain moisture (NIEDER et al., 1999; ACEBEY; KROMER, 2001), which is very important for the vascular epiphyte community in dry forest areas such as the one found at this site. However, the low abundance of the low stem, together with the lack of a significant difference in the richness of this stem and the outer canopy, suggest that this forest remnant is altered in terms of its conservation status.

Analysis of the vascular epiphytes at Sitio Réplica II in the Central Area of the Sorocaba/Médio Tietê river basin

The Replica II site of Semideciduous Seasonal Forest in the central area of the watershed was established in a legal reserve area of a condominium in the municipality of Porto Feliz - SP. The forest fragment located between coordinates UTM 251.231 and 7.416.613 in zone 23 South is a remnant of Semideciduous Seasonal Forest and has an area of approximately 104 ha, making it the largest forest fragment in the municipality. Although part of the fragment is protected by law, most of the area is under heavy anthropogenic interference, including cattle and horse breeding in the forest remnant's understory. This means that there is practically no forest regeneration or tree succession, which is fundamental for the development and maintenance of the epiphytic vascular community, which is therefore compromised (Figure 27).

Figura 27:   Aspect of the understory of Sitio Règlica II in the central area of the watershed.

At Sitio Réplica II in the central area of the Sorocaba/Médio Tietê basin, 14 species were found, belonging to eight genera and four families (Table 26). The site's Shannon index was H' = 1.695, equability (J) was 0.642 and Margalef's richness (d) was 1.997.

Table 26 - List of vascular epiphyte species found at Sitio Réplica II in the central area (Condominio FARM) of the Sorocaba Médio Tietê hydrographic basin and their respective Ecological Categories (EC) - HLC: Holoepiphyte characteristic; HLF: Holoepiphyte facultative; HLA: Holoepiphyte accidental; HMP: Hemiepiphyte primary: HMS: Hemiepiphyte secondary. Dispersal form (Disp.) - Zo: = Zoochoric; An: Anemochoric. Reg: HUFSCar herbarium registration number

N	Family	Species	EC	Disp.	Reg.
1	Bromeliaceae	Billbergia distachya (Vell.) Mez	HLC	Zo	8445
2	Bromeliaceae	Tillandsia recurvata (L.) L.	HLC	An	8429
3	Bromeliaceae	Tillandsia stricta Sol. ex Sims	HLC	An	8428
4	Cactaceae	Epiphyllum phyllanthus (L.) Haw.	HLC	Zo	8474
5	Cactaceae	Lepismium lumbricoides (Lem.) Barthlott	HLC	Zo	8386
6	Orchidaceae	Polystachya foliosa (Lindl.) Rchb.f.	HLC	An	8471
7	Orchidaceae	Encyclia oncidioides (Lindl.) Schltr.	HLC	An	8466
8	Polypodiaceae	Microgramma persicariifolia (Schrad.) C. Presl	HLC	An	8520
9	Polypodiaceae	Microgramma vacciniifolia (Langsd. & Fisch.) Copel.	HLC	An	8522
10	Polypodiaceae	Pleopeltis astrolepis (Liebm.) E. Fourn.	HLC	An	8519
11	Polypodiaceae	Pleopeltis hirsutissima (Raddi) de la Sota	HLC	An	8521
12	Polypodiaceae	Pleopeltis pleopeltifolia (Raddi) Alston	HLC	An	8525
13	Polypodiaceae	Pleopeltis squalida (Vell.) de la Sota	HLC	An	8436
14	Polypodiaceae	Serpocaulon latipes (Langsd. & L. Fisch.) A.R. Sm.	HLF	An	8423

The richness of the area can be considered low when compared to studies carried out in the same type of forest, e.g. the studies by Rogalski and Zanin (2003) - 70 species; Giongo and Waechter (2004) - 57 species; Cervi and Borgo (2007) - 56 species; Dislich and Mantovani (1998) - 34 species; Borgo et al. (2002) - 32 species; Breier (2005) - 25 species and Dettke et al. (2008) - 29 species. This site showed lower richness than that found in the other sites (Core and Replicas) sampled in the Central Area of this basin. The low number of epiphytic species may be related to the lack of complexity of the forest (especially given the absence of understorey, which has been suppressed by cattle ranching), either by the absence of a favorable microclimate or even by the reduction in the number of forophytes (especially larger trees), since these factors may be responsible for reducing epiphytic diversity (ENGWALD et al., 2000; BARTHLOTT et al., 2001; DETTKE et al., 2008).

The Polypodiaceae family, cited by Dettke et al. (2008) as the richest in an altered fragment of Semideciduous Seasonal Forest in the city of Maringà, was responsible for 50% of the epiphytic species in the area (seven species), followed by the Bromeliaceae with three species (21%), which is also cited by the above authors. The Cactaceae and Orchidaceae families had two species each. Characteristic holoepiphytes predominated at this site, accounting for almost 93% of the species, followed by facultative holoepiphytes with 7.1% of the species. No accidental holoepiphytes, primary hemiepiphytes or secondary hemiepiphytes were recorded. Among the epiphytic species, 11 (79%) showed anemochoric dispersal and only three (21%) showed zoochoric dispersal. The predominance of anemochory has been reported in other studies carried out in Brazil (BREIER, 2005; DETTKE et al., 2008; MENINI-NETO et al., 2009; GERALDINO et al., 2010).

In the quantitative analysis, 14 species were recorded, with three Polypodiaceae being the most prominent species at Sitio Réplica II (Table 27). *Microgramma Vacciniifolia* had an epiphytic importance value (VIE) of 45.17 and an average score of 2.06, occurring in 90% of the forophytes and 51.3% of the strata, making it the most important species at this site. *Pleopeltis pleopeltifolia* had a VIE of 22.98, an average score of 1.58 and was recorded in 66.7% of the forophytes and 34.1% of the strata.

Table 27 - Vascular epiphytes from Sitio Rèplica II in the central area of the Sorocaba Médio Tietê hydrographic basin, classified according to epiphytic importance value - nr: absolute number of occurrences in the strata; far: absolute frequency in the strata; ni: absolute number of occurrences in the forophytic individuals; fai: absolute frequency in the forophytic individuals; vt (total value): sum of the abundance estimates; vie: epiphytic importance value; nota: average score obtained

Species	nr	far	ni	fai	vt	vie	note
Microgramma vacciniifolia	277	51,3	81	90,0	570	45,17	2,06
Pleopeltis pleopeltifolia	184	34,1	60	66,7	290	22,98	1,58
Pleopeltis squalida	46	8,5	16	17,8	105	8,32	2,28
Polystachya foliosa	38	7,0	17	18,9	87	6,89	2,29
Pleopeltis hirsutissima	42	7,8	21	23,3	73	5,78	1,74
Tillandsia recurvata	39	7,2	15	16,7	59	4,68	1,51
Epiphyllum phyllanthus	19	3,5	11	12,2	34	2,69	1,79
Tillandsia stricta	11	2,0	5	5,6	16	1,27	1,45
Serpocaulon latipes	5	0,9	3	3,3	10	0,79	2,00
Billbergia distachya	4	0,7	3	3,3	8	0,63	2,00
Microgramma persicariifolia	2	0,4	1	1,1	3	0,24	1,50
Pleopeltis astrolepis	2	0,4	1	1,1	3	0,24	1,50

| Lepismium lumbricoides | 1 | 0,2 | 1 | 1,1 | 2 | 0,16 | 2,00 |
| *Encyclia oncidioides* | 1 | 0,2 | 1 | 1,1 | 2 | 0,16 | 2,00 |

Pleopeltis squalida, another Polypodiaceae, recorded in 17.8% of the forophytes and 8.5% of the strata, with an average score of 2.28, had a VIE of 8.32 and was the third most important species at Sitio Règlica II. These three species were responsible for more than 76% of the VIE of this site. The Polypodiaceae family, which had the highest richness in the area, was also responsible for almost 84% of the epiphytic importance value. The Orchidaceae family (Figure 28), although it only had two species, was the second most important in the site with a VIE of 7.05. The concentration of abundance in a few species is typical of impacted areas (BATAGHIN et al., 2010) and may be related to models of pre-empting niches (MAY, 1975). At this site, the tendency for species to be concentrated in a few families is repeated, and these are the predominant species in areas under anthropogenic pressure. One fact that stands out and serves as an indication of the conservation status of the forest remnant is the presence of a species of Orchidaceae from the genus Polystachya among the most abundant species on the site. This genus is commonly found in forests that have been altered or are at an intermediate sectional stage.

Figura 28:   Detail of Encyclia oncidioides (Lindl.) Schltr. (Orchidaceae) present at Sitio Règlica II in the Central Area of the hydrographic basin.

The distribution of epiphytes in the forophyte strata (Figure 29) showed that the base of the canopy, with an abundance value (VA) of 437, was the stratum with the highest epiphyte abundance. The second most abundant stratum was the inner canopy with a VA of 366, followed by the high stem with a VA of 255, the middle stem with a VA of 103, the outer canopy with a VA of 74 and the low stem with a VA of 27.

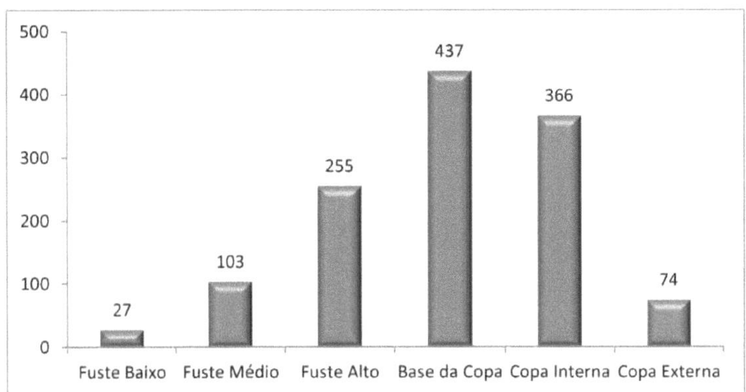

Figura 29:   Distribution of the abundance of epiphytic vascular species among the forophytic strata at Sitio Rèplica II in the central area of the Sorocaba/Médio Tietê hydrographic basin.

The vertical distribution of the vascular epiphytes on the forophytes based on their abundance revealed that the base of the canopy had the highest number of individuals, and this stratum differed significantly from the low and medium stems and also from the outer canopy (Table 28). The low stem was also significantly different from the high stem and the inner canopy; this stem (low) had a very small number of individuals (Figure 29), which is due to the environmental conditions not favorable to epiphytism, which in turn can be attributed to the lack of structural complexity in this stratum, given that the understory was practically eliminated in this site.

Table 28: Analysis of the distribution of the richness and abundance of vascular epiphytes among the forophytic strata at Sitio Rèplica II in the Central Area of the Sorocaba/Médio Tietê basin.

^^^-^^Richness Abundance"---^	Beam Bass	Beam Medium	Beam High	World base	Cup Cup Internal	Cup External
Low stem		0,7297	0,7297	0,05018	0,2658	0,4695
Medium stem	0,08		0,9786	0,1066	0,4551	0,2758
High stem	**0,033**	0,124		0,1066	0,4551	0,2758
World Cup base	**0,02**	**0,049**	0,21		0,3831	**0,0085**
Internal Cup	**0,043**	0,095	0,311	0,396		0,0666
External Cup	0,141	0,329	0,08	**0,035**	0,071	

In terms of richness, the epiphytic community that develops at Sitio Réplica II in the Central Area is simple, and so strongly limited by microclimatic factors that practically the same species occur from the canopy to the ground, with only the number of individuals varying. The only exception is the base of the canopy, which showed a significant difference from the outer canopy. In the intermediate regions of the forophytes, some exclusive species are recorded, such as Encyclia oncidioides (Figure 28), which is more demanding in terms of living environment. In general for this site (with a strong anthropogenic influence), what varied was the number of individuals within the species, but the communities that develop in each stratum are similar.

Analysis of the vascular epiphytes of Sitio Rèplica III in the Central Area of the Sorocaba/Médio Tietê basin

At Sitio Rèplica III in the central area of the Sorocaba/Médio Tiete river basin, the survey of vascular epiphytes was carried out on a private rural property in the municipality of Araçoiaba da Serra - SP. The forest remnant covers around 31 ha and is located on the banks of the Iperó River, between UTM coordinates 228.738 and 7.398.832 in zone 23 South, and is characterized by being a stretch of Semideciduous Seasonal Forest. As with the other sites in the central area, there are also signs of anthropogenic influence at Sitio Réplica III, such as the presence of trails. However, in this area there is an abundant understorey which contributes to maintaining the microclimate (Figure 30). The surrounding areas are characterized by agricultural use and cattle breeding.

Figura 30: Aspect of the understory of Sitio Réplica III in the Central Area of the Sorocaba/Médio Tietê river basin.

Sitio Réplica III was the area with the highest diversity of epiphytic species; 33 species were recorded, belonging to 19 genera and eight families (Table 29). The Shannon index for the area was H' = 2.564, the equability (J) was 0.739 and the Margalef richness (d) was 4.58.

Table 29 - List of vascular epiphyte species found at Sitio Réplica III in the central area (Fazenda Sao José) of the Sorocaba Médio Tietê hydrographic basin and their respective Ecological Categories (EC) - HLC: Holoepiphyte characteristic; HLF: Holoepiphyte facultative; HLA: Holoepiphyte accidental; HMP: Hemiepiphyte primary: HMS: Hemiepiphyte secondary. Dispersal form (Disp.) - Zo: Zoochoric; An: Anemochoric. Reg: HUFSCar herbarium registration number (Im: Digital image).

N	Family	Species	EC	Disp.	Reg.
1	Araceae	Anthurium comtum Schott	HLF	Zo	8486
2	Araceae	Philodendron bipinnatifidum Schott	HMP	Zo	Im
3	Araceae	Philodendron eximium Schott	HMP	Zo	8488
4	Bromeliaceae	Aechmea bromeliifolia (Rudge) Baker	HLC	Zo	8432
5	Bromeliaceae	Aechmea nudicaulis (L.) Griseb.	HLF	Zo	8434

Keep going...

Table 29 - Continuation...

N	Family	Species	EC	Disp.	Reg.
6	Bromeliaceae	*Billbergia distachya* (Vell.) Mez	HLC	Zo	8445
7	Bromeliaceae	*Tillandsia funckiana* Baker	HLC	An	8458
8	Bromeliaceae	*Tillandsia recurvata* (L.) L.	HLC	An	8429
9	Bromeliaceae	*Tillandsia stricta* Sol. ex Sims	HLC	An	8428
10	Cactaceae	*Epiphyllum phyllanthus* (L.) Haw.	HLC	Zo	8474
11	Cactaceae	*Lepismium cruciforme* (Vell.) Miq.	HLC	Zo	8388
12	Cactaceae	*Lepismium lumbricoides* (Lem.) Barthlott	HLC	Zo	8386
13	Cactaceae	*Rhipsalis pilocarpa* Loefgr.	HLC	Zo	8385
14	Cactaceae	*Rhipsalis teres* (Vell.) Steud.	HLC	Zo	8384
15	Commelinaceae	*Tradescantia albiflora* Kunth	HLA	Zo	8391
16	Orchidaceae	*Cyclopogon multiflorus* Schltr.	HLA	An	8404
17	Orchidaceae	*Epidendrum rigidum* Jacq.	HLC	An	8453
18	Orchidaceae	*Lophiaris pumila* (Lindl.) Braem	HLC	An	8468
19	Piperaceae	*Peperomia trineuroides* Dahlst.	HLC	Zo	8450
20	Piperaceae	*Peperomia glabella* (Sw.) A. Dietr.	HMP	Zo	8475
21	Piperaceae	*Peperomia pereskiifolia* (Jacq.) Kunth	HMP	Zo	8448
22	Piperaceae	*Peperomia tetraphylla* (G. Forst.) Hook. & Arn.	HLC	Zo	8447
23	Polypodiaceae	*Campyloneurum centrobrasilianum* Lellinger	HLC	An	8526
24	Polypodiaceae	*Campyloneurum nitidum* (Kaulf.) C. Presl	HLC	An	8531
25	Polypodiaceae	*Microgramma persicariifolia* (Schrad.) C. Presl	HLC	An	8520
26	Polypodiaceae	*Microgramma tecta* (Kaulf.) Alston	HLC	An	8524
27	Polypodiaceae	*Microgramma vacciniifolia* (Langsd. & Fisch.) Copel.	HLC	An	8522
28	Polypodiaceae	*Pecluma filicula* (Kaulf.) M.G. Price	HLC	An	8527

29	Polypodiaceae	*Pleopeltis astrolepis* (Liebm.) E. Fourn.	HLC	An	8519
30	Polypodiaceae	*Pleopeltis hirsutissima* (Raddi) de la Sota	HLC	An	8521
31	Polypodiaceae	*Pleopeltis pleopeltifolia* (Raddi) Alston	HLC	An	8525
32	Polypodiaceae	*Serpocaulon latipes* (Langsd. & L. Fisch.) A.R. Sm.	HLF	An	8423
33	Pteridaceae	*Vittaria lineata* (L.) Sm.	HLC	An	8454

Considering other studies carried out in Semideciduous Seasonal Forest, the richness of Sitio Règlica III was similar to or higher than the data obtained by Aguiar et al. (1981), who sampled 17 species; by Dislich and Mantovani (1998), 34 species; by Borgo et al. (2002), with 32 species; by Breier (2005), 25 species; by Dettke et al. (2008), 29 species and Bataghin et al. (2010), 21 species, although it was lower than the results of Rogalski and Zanin (2003) who found 70 species, Giongo and Waechter (2004) who sampled 57 species and Cervi and Borgo (2007), who found 56 species.

Sitio Règlica III, despite having the smallest area (31 ha) of all the sites studied in the Central Area, had the highest diversity of vascular epiphytes. Although it is common sense and has been reported by several studies that changes in the landscape have a negative influence on the diversity and abundance of vascular epiphytes (ENGWALD et al., 2000; BONNET; QUEIROZ, 2000; BARTHLOTT et al., 2001; BATAGHIN et al.., 2008; DETTKE et al. 2008), small forest remnants should not be ignored or left aside in conservation actions, as it seems that forest fragmentation may have isolated or even restricted the epiphytic community in these forest fragments. This also highlights the importance of conserving forest areas, even impacted ones, or even isolated tree individuals in the landscape, for the conservation of the vascular epiphytic community (BARTHLOTT et al., 2001; DETTKE et al., 2008; KERSTEN; KUNIYOSHI, 2009; BATAGHIN et al., 2010).

The richest family at Sitio Règlica III was Polypodiaceae with 10 species. The Bromeliaceae and Cactaceae families had six and five species, respectively. They were followed by Piperaceae (four species), Araceae and Orchidaceae (three species each). The Commelinaceae and Pteridaceae families had one epiphytic species each.

Characteristic holoepiphytes were dominant at the site, accounting for 23 species (72%), followed by facultative holoepiphytes and primary hemiepiphytes, both with four species (13%) each. Accidental holoepiphytes contributed two species and no secondary hemiepiphytes were recorded. The predominance of characteristic holoepiphytes observed has been common in research carried out in areas of Semideciduous Seasonal Forest (DISLICH; MANTOVANI, 1998; BORGO et al., 2002, ROGALSKI; ZANIN, 2003; KERSTEN; KUNIYOSHI, 2009; BATAGHIN et al., 2010). As for the dispersal syndrome, 51.5% of the species (17 spp.) were anemochoric and 48.5% (16 spp.) were zoochoric. There seems to be a tendency for the expected proportion (2/3) of anemochoric species to be reduced in isolated forest fragments with a small area or with a small number of large trees, as is the case at this site.

Of the 33 species found in the floristic survey, 32 were recorded in the quantitative analysis of Sitio Réplica III (Table 30).

Table 30 - Vascular epiphytes from Sitio Rèplica III in the central area of the Sorocaba Médio Tietê river basin, classified according to epiphytic importance value - nr: absolute number of occurrences in the strata; far: absolute frequency in the strata; ni: absolute number of occurrences in the forophytic individuals; fai: absolute frequency in the forophytic individuals; vt (total value): sum of the abundance estimates; vie: epiphytic importance value; nota: average score obtained.

Species	nr	far	ni	fai	vt	vie	note
Microgramma vacciniifolia	178	33,0	53	58,9	331	19,73	1,86
Pleopeltis pleopeltifolia	147	27,2	47	52,2	285	16,98	1,94
Microgramma tecta	129	23,9	38	42,2	237	14,12	1,84
Billbergia distachya	81	15,0	45	50,0	150	8,94	1,85
Rhipsalis teres	57	10,6	25	27,8	117	6,97	2,05
Lophiaris pumila	58	10,7	34	37,8	112	6,67	1,93
Aechmea bromeliifolia	23	4,3	16	17,8	56	3,34	2,43
Microgramma persicariifolia	22	4,1	8	8,9	45	2,68	2,05
Aechmea nudicaulis	19	3,5	12	13,3	43	2,56	2,26
Lepismium cruciforme	16	3,0	6	6,7	39	2,32	2,44
Serpocaulon latipes	18	3,3	8	8,9	36	2,15	2,00
Tillandsia recurvata	22	4,1	10	11,1	30	1,79	1,36
Epiphyllum phyllanthus	13	2,4	7	7,8	29	1,73	2,23
Pleopeltis hirsutissima	13	2,4	7	7,8	28	1,67	2,15
Peperomia glabella	14	2,6	9	10,0	26	1,55	1,86
Campyloneurum nitidum	12	2,2	7	7,8	21	1,25	1,75
Peperomia tetraphylla	7	1,3	2	2,2	15	0,89	2,14
Anthurium comtum	5	0,9	2	2,2	9	0,54	1,80
Tillandsia stricta	5	0,9	4	4,4	9	0,54	1,80
Peperomia pereskiifolia	3	0,6	2	2,2	8	0,48	2,67
Philodendron bipinnatifidum	3	0,6	2	2,2	8	0,48	2,67
Vittaria lineata	4	0,7	2	2,2	7	0,42	1,75
Campyloneurum centrobrasilianum	3	0,6	3	3,3	6	0,36	2,00
Tillandsia funckiana	4	0,7	3	3,3	6	0,36	1,50
Epidendrum cf. *rigidum*	3	0,6	2	2,2	5	0,30	1,67
Pecluma filicula	2	0,4	1	1,1	4	0,24	2,00
Peperomia trineuroides	2	0,4	1	1,1	4	0,24	2,00

Pleopeltis astrolepis	2	0,4	2	2,2	4	0,24	2,00
Lepismium lumbricoides	2	0,4	1	1,1	3	0,18	1,50
Philodendron eximium	1	0,2	1	1,1	2	0,12	2,00
Tradescantia albiflora	1	0,2	1	1,1	2	0,12	2,00
Rhipsalis pilocarpa	1	0,2	1	1,1	1	0,06	1,00

Three Polypodiaceae were the most prominent species at this site. *Microgramma Vacciniifolia* (Figure 31) had an epiphytic importance value (EVI) of 19.73 and an average score of 1.86, and was present in almost 59% of the forophytes and 33% of the strata, making it the most important species at this site. *Pleopeltis pleopeltifolia* was the second most important species with a VIE of 16.98 and an average score of 1.94, as well as being present in 52.2% of the forophytes and 27.2% of the strata. The third most important species at this site was *Microgramma tecta,* with a VIE of 14.12 and a score of 1.84, which was recorded in over 42% of the forophytes and almost 24% of the strata.

Figura 31: Detail of *Microgramma vacciniifolia* (Langsd. & Fisch.) Copel. (Polypodiaceae), the species with the highest epiphytic importance value at Sitio Règlica III in the Central Area.

The three species mentioned above contributed to making the Polypodiaceae family the most important in Sitio Règlica III with almost 60% of the epiphytic importance value. The Bromeliaceae family was the second most important with a total VIE of 17.52, followed by Cactaceae with a VIE of 11.26 and Orchidaceae with a VIE of 6.97. Although the Polypodiaceae and Bromeliaceae - significant families in areas under anthropogenic pressure according to Dettke et al. (2008) - are predominant in the site, the greater diversity of species compared to the other sites studied in this research, as well as the outstanding epiphytic importance value of the Orchidaceae family (although with only three species), indicate a favorable environment for the establishment and development of vascular epiphytes in this site.

With regard to the vertical distribution of epiphytes, the base of the canopy (Figure 32) was the most abundant forophytic stratum with an abundance value (AV) of 571, followed by the inner canopy, the second most abundant stratum with an AV of 374, and the high stem with an AV of 356. The middle canopy had an abundance value of 200, *while the* outer canopy and low canopy had abundance values of 103 and 74, respectively.

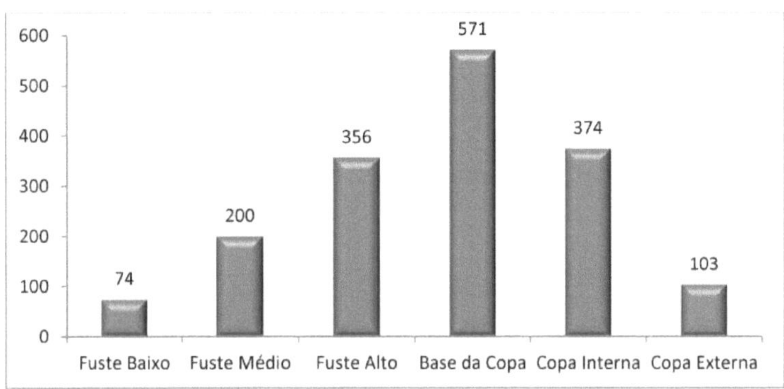

Figura 32: Distribution of the abundance of epiphytic vascular species among the forophytic strata at Sitio Réplica III in the central area of the Sorocaba/Médio Tietê hydrographic basin.

The statistical analysis applied to the distribution of vascular epiphyte abundances in the forophytic strata showed that the low stem differed from all strata except the outer canopy (Table 31). The middle stem differed only from the base of the canopy. Still in terms of abundance, the outer canopy showed significant differences from the other three strata, but its richness was different from all the other strata in this site.

Table 31: Analysis of the distribution of the richness and abundance of vascular epiphytes among the forophytic strata at Sitio Réplica III in the Central Area of the Sorocaba/Médio Tietê basin.

Wealth Abundance^^^^	Beam Bass	Beam Medium	Beam High	Cup base	Cup Internal	Cup External
Low stem		0,0529	0,199	0,199	0,8078	**2,65E-04**
Medium stem	**0,007**		0,3848	0,3848	0,0585	**4,77E-08**
High stem	**0,002**	0,067		0,9931	0,3001	**1,90E-06**
Cup base	**0,001**	**0,014**	0,122		0,3001	**1,90E-06**
Internal Cup	**0,025**	0,134	0,459	0,184		**1,09E-04**
External Cup	0,316	0,094	**0,011**	**0,004**	**0,047**	0

The lower abundances of the low canopy and the outer canopy were already expected and can be attributed to the limitations encountered by the epiphytic vascular community in developing in these extreme locations, especially for the acquisition of water (outer canopy) and light (low canopy). Likewise, the greater abundance of intermediate strata was expected (STEEGE; CORNELISSEN, 1989; FREIBERG, 1996). What is striking about this site is the richness of the forophyte strata, where once the zone of greatest climatic fluctuation (outer canopy) has been overcome, the epiphytic community tends to be more homogeneous in terms of species composition. This site, although small in size (around 30 ha), has excellent internal conditions, with an abundant understorey and the presence of larger trees. These factors seem to have a great influence on the development of epiphytes, although a more detailed study of the structural complexity of forest fragments could shed more light on this. In addition, in the (relatively) close vicinity of this forest remnant there are other small forest fragments, which may contribute to the dispersal of both anemochoric and zoochoric epiphytic species, since numerically they are almost equivalent at this site. Again, there is a need for further studies involving vascular epiphytes on a larger scale, possibly on a landscape scale.

Analysis of the vascular epiphytes of the Qualitative Sites (QI, QII, QIII) in the Central Area of the Sorocaba/Médio Tietê basin

The survey of vascular epiphytes in Qualitative Sites QI, QII and QIII in the Central Area of the Sorocaba/Médio Tiete river basin took place in the following forest fragments. Qualitative Site I (QI) was established in an area of approximately 30 ha, located between the UTM coordinates 199.635 and 7.436.241 of zone 23 south in the municipality of Pereiras - SP. The IQ's phytophysiognomy is characterized as Semideciduous Seasonal Forest and its surroundings are predominantly agricultural, with cattle breeding, as well as an area with sugar cane cultivation. The floristic analysis of Qualitative Site II (QII) took place in the municipality of Cesàrio Lange - SP in a fragment of Semideciduous Seasonal Forest of approximately 70 ha located between coordinates UTM 200.835 and 7.433.920 in zone 23 south. The forest fragment's surroundings are characterized by the presence of pastures for cattle breeding. Also noteworthy is the presence of trails within the forest formation which are clearly used routinely for hiking. Qualitative Site III (QIII), located between UTM coordinates 204.110 and 7.437.847 in zone 23 South, is a 97.3 ha forest fragment located inside a sugar cane farm in the municipality of Laranjal Paulista, and can be characterized as an area of Semideciduous Seasonal Forest. Although all three qualitative sites showed signs of anthropic influence, such as the presence of trails, it is worth noting that only Site QII showed clear evidence of constant human presence (Figure 33), while Qualitative Sites QI and QIII are better preserved.

Figura 33:   Existing trail (road) within the Qualitative Site II of the Central Area.

In the three Qualitative Sites (QI, QII, QIII) in the Central Area of the Sorocaba/Médio Tietè river basin, 38 species were recorded, belonging to 22 genera and seven families (Table 32).

Table 32 - List of the vascular epiphyte species found at the Qualitative Sites in the Central Area of the Sorocaba Mèdio Tietè river basin and their respective Ecological Categories (EC) - HLC: Characteristic Holoepiphyte; HLF: Facultative Holoepiphyte; HLA: Accidental Holoepiphyte; HMP: Primary Hemiepiphyte: HMS: Secondary Hemiepiphyte. Dispersal form (Disp.) - Zo: = Zoochoric; An: Anemochoric. Q: Qualitative Survey Site (1, 2, 3).

N	Family	Species	EC	Disp.	Sitio
1	Araceae	Philodendron bipinnatifidum Schott	HMS	Zo	Q1; Q2; Q3;
2	Bromeliaceae	Acanthostachys strobilacea (Schult. f.) Klotzsch	HLC	Zo	Q1; Q3;
3		Aechmea distichantha Lem.	HLF	Zo	Q2;

Table 32 - Continuation...

101

N	Family	Species	EC	Disp.	Sitio
4		*Billbergia amoena* (Lodd.) Lindl.	HLC	Zo	Q2;
5		*Billbergia distachya* (Vell.) Mez	HLC	Zo	Q1;
6		*Billbergia zebrina* (Herb.) Lindl.	HLC	Zo	Q2; Q3
7		*Tillandsia recurvata* (L.) L.	HLC	An	Q1; Q2; Q3;
8		*Tillandsia* sp.	HLC	An	Q1;
9		*Tillandsia stricta* Sol. ex Sims	HLC	An	Q1; Q2; Q3;
10		*Tillandsia tricholepis* Baker	HLC	An	Q1; Q2; Q3;
11		*Tillandsia usneoides* (L.) L.	HLC	An	Q1; Q3;
12		*Vriesea fenestralis* Linden & André	HLC	An	Q1;
13		*Vriesea friburgensis* Mez	HLF	An	Q2;
14		*Vriesea procera* (Mart. ex Schult. & Schult.f.) Wittm.	HLF	An	Q1;
15	Cactaceae	*Cereus alacriportanus* Pfeiff.	HLF	Zo	Q1; Q3;
16		*Epiphyllum phyllanthus* (L.) Haw.	HLC	Zo	Q2; Q3
17		*Lepismium cruciforme* (Vell.) Miq.	HLC	Zo	Q1; Q2; Q3;
18		*Lepismium lumbricoides* (Lem.) Barthlott	HLC	Zo	Q2;
19		*Rhipsalis cereuscula* Haw.	HLC	Zo	Q1;
20		*Rhipsalis floccosa* Salm-Dyck ex Pfeiff.	HLC	Zo	Q2;
21		*Rhipsalis teres* (Vell.) Steud.	HLC	Zo	Q1; Q2; Q3;
22		*Rhipsalis trigona* Pfeiff.	HLC	Zo	Q1; Q2; Q3;
23	Orchidaceae	*Baptistonia lietzei* (Regel) Chiron & V.P.Castro	HLC	An	Q1;
24		*Brasiliorchis chrysantha* (Barb. Rodr.) R.B.Singer *et al.*	HLC	An	Q1; Q3;
25		*Miltonia flavescens* (Lindl.) Lindl.	HLC	An	Q1;
26		*Polystachya estrellensis* Rchb.f	HLC	An	Q2;
27		*Rodriguezia decora* (Lem.) Rchb. f.	HLC	An	Q1; Q2; Q3;
28		*Sophronitis cernua* Lindl.	HLC	An	Q3;
29	Piperaceae	*Peperomia tetraphylla* (G. Forst.) Hook. & Arn.	HLC	Zo	Q1; Q2;
30	Polypodiaceae	*Campyloneurum repens* (Aubl.) C. Presl	HLC	An	Q1; Q2;
31		*Microgramma persicariifolia* (Schrad.) C. Presl	HLC	An	Q2; Q3
32		*Microgramma tecta* (Kaulf.) Alston	HLC	An	Q2;
33		*Microgramma squamulosa* (Kaulf.) de la Sota	HLC	An	Q1; Q2; Q3;
34		*Pecluma filicula* (Kaulf.) M.G. Price	HLC	An	Q1;
35		*Pleopeltis hirsutissima* (Raddi) de la Sota	HLC	An	Q1;
36		*Pleopeltis pleopeltifolia* (Raddi) Alston	HLC	An	Q1; Q2; Q3;
37		*Pleopeltis squalida* (Vell.) de la Sota	HLC	An	Q1; Q2; Q3;
38	Pteridaceae	*Polytaenium cajenense* (Desv.) Benedict	HLC	An	Q2;

At Sitio QI, 26 species belonging to 17 genera and six families were recorded. Characteristic holoepiphytes were predominant at this site (23 species), followed by facultative holoepiphytes with two species and primary hemiepiphytes with one species. No accidental holoepiphytes or secondary hemiepiphytes were recorded at Sitio QI. The Bromeliaceae family was the richest at this site with nine species, followed by Polypodiaceae and Cactaceae with six and five species, respectively. Orchidaceae had four species and the Araceae and Piperaceae families had one species each. Site QI was the richest of the qualitative sites in the Central Area of this watershed.

At Sitio QII in the central area, there were 24 species belonging to 15 genera and seven families. The characteristic holoepiphytes accounted for 20 species, followed by facultative holoepiphytes with three and primary hemiepiphytes with one species. Bromeliaceae, with seven species, was dominant at this site, followed by Polypodiaceae and Cactaceae with six species each. Orchidaceae accounted for two species and the Araceae, Piperaceae and Pteridaceae families had one species each.

Site QIII was the least diverse of the qualitative sites in the Central Area, with 19 species belonging to 13 genera and five families. The characteristic holoepiphytes were responsible for 17 species, and the facultative holoepiphytes and primary hemiepiphytes had one species each. Bromeliaceae with six species was predominant at this site, followed by Cactaceae with five species and Polypodiaceae with four species. Orchidaceae accounted for three species and the Araceae family had one species.

The richness of the Qualitative Sites (QI, QII, QIII), analyzed together here, is low when compared to the work of Rogalski and Zanin (2003) who found 70 species, Giongo and Waechter (2004) who sampled 57 species, Cervi and Borgo (2007) who found 56 species, Menini-Neto et al. (2009) with 59 species (Descoberto - MG), Alves and Kolbek 2009 who sampled 61 species and Bonnet et al. (2011) who also sampled 61 species. However, it is similar to the studies carried out in Semideciduous Seasonal Forest by Dislich and Mantovani (1998), with 34 species; by Borgo et al. (2002), with 32 species and by Menini-Neto et al. (2009) with 41 species (Barroso - MG). In addition to being superior to the results obtained by Aguiar et al. (1981), who sampled 17 species; by Breier (2005), with 25 species, by Dettke et al. (2008), with 29 species and by Bataghin et al. (2010), with 21 species.

The richest families in the Qualitative Sites (QI, QII, QIII) in the Central Area were Bromeliaceae (Figure 34) with 13 species, and Polypodiaceae and Cactaceae with eight species each. The Orchidaceae family had six species. Araceae, Piperaceae and Pteridaceae had one species each.

Figura 34:   Billbergia zebrina (Herb.) Lindl. (Bromeliaceae) recorded at two of the three Qualitative Sites in the Central Area.

In general, the characteristic holoepiphytes were dominant at the site, accounting for 32 species (84%), followed by facultative holoepiphytes with five species (13%), and primary hemiepiphytes with one species. No accidental holoepiphytes or secondary hemiepiphytes were recorded at the qualitative sites in the Central Area.

Of all the species, 23 (60%) showed anemochoric dispersal and 15 (40%) zoochoric dispersal. This result partly reflects that observed by (Benzing 1987), who recorded 2/3 of the species as anemochoric. However, the reduction in the proportion of anemochoric/zoochoric species provides an indication that small anthropically altered forest fragments or those of low structural complexity (observed in the qualitative sites in this area) tend to increase the number of epiphytic species with a zoochoric dispersal syndrome.

Analysis of the vascular epiphytes of the Seasonal Semideciduous Forest in the Central Area of the Sorocaba/Médio Tietê Hydrographic Basin

The analysis of the distribution of species abundances between the sites showed no significant differences between the Core Site and its replicas; the values obtained were as follows: (a) the analysis between the Core Site and Replica Site I showed t = 1.173 and p = 0.122; (b) between the Core Sites and Replica II, t = 0.183 and p = 0.428; and (c) between the Core Sites and Replica III, t = 0.400 and p = 0.345. The absence of a significant difference between the abundance of the Core Site and its replicas is an indication that the vascular epiphytic community that develops in the Semideciduous Seasonal Forest of the central area of the Sorocaba/Médio Tietê river basin is similar in terms of the number of individuals in the different sites studied.

The same statistical analysis applied to the presence/absence of species showed that the Core Site differed only from the Replica II and Replica III sites; the other sites showed no significant difference from the Core Site. The analysis yielded the following results: (a) between Sites Core and Replica I, $t = 0.729$ and $p = 0.234$; (b) between Site Core and Site Replica II, $t = 2.239$ and $p = 0.014$; (c) between Sites Core and Replica III, $t = -2.167$ and $p = 0.016$; (d) between Core Sites and QI, $t = -0.712$ and $p = 0.239$; (e) between Core Sites and QII, $t = -0.251$ and $p = 0.401$; and (f) between Core Sites and QIII, $t = 1.045$ and $p = 0.15$. In general, the analysis shows that the diversity of species found at the Core Site in the Central Area is similar to most of the surrounding sites. However, when comparing the diversity of the Core Site (a Conservation Unit - 23 species) with the total diversity of the other sites (unprotected forest fragments in the surrounding area - 61 species), a significant difference was found with $t = -8.517$ and $p = 0.0001$. This result highlights the importance of conserving forest fragments for vascular epiphytic diversity, whether these fragments are protected by Conservation Units or not.

The Jaccard Similarity Index, which can be seen in Figure 35, showed similarities of: 30% between Sitio Core and Sitio Réplica I; 19% between Sitio Core and Sitio Réplica II; 27% between Sitio Core and Sitio Réplica III; 28% between Sitio Core and Sitio QI; 34% between Sitio Core and Sitio QII; and 35% between Sitio Core and Sitio QIII. The greatest floristic similarity occurred between Sitio QI and Sitio QIII, which shared 50% of the species. This greater similarity between Sites QI and QIII can be attributed to the geographical proximity of the forest fragments, which is reflected in similar climatic and vegetational characteristics.

The low similarity between the sites corroborates the idea that the vascular epiphytic community found in the Central Area of the Sorocaba/Médio Tietê basin is heterogeneous in terms of species composition, revealing the importance of forest fragments not protected by UCs in vascular epiphytic diversity.

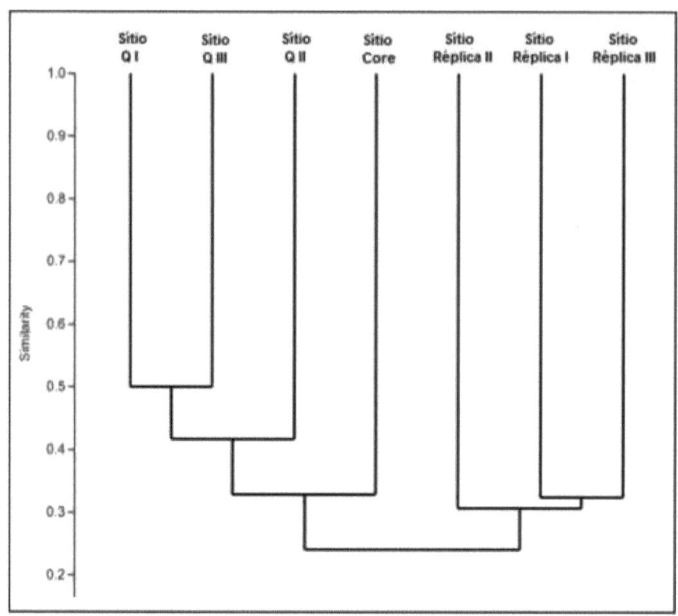

Figura 35: Dendrogram (UPGMA) of Jaccard's similarity between sites in the Semideciduous Seasonal Forest of the Central Area of the Sorocaba/Médio Tietê basin.

It is interesting to note the low level of similarity between the vascular epiphytes sampled in the Central Area of the Sorocaba/Médio Tietê basin and other studies carried out on vascular epiphytes in different forest formations, both in areas of Semideciduous Seasonal Forest and Cerrado areas in Brazil (Table 33).

Table 33 - Jaccard's similarity index (IJ) between the vascular epiphytic species from the Central Area of the Sorocaba/Médio Tietê basin and other studies of vascular epiphytes carried out in Semideciduous Seasonal Forest in Brazil.

Source	Location (Phytophysiognomy)	Species	Families	Jaccard (J)
Aguiar et al. 1981	Montenegro - RS (FES)	17	5	0,161
Dislich and Mantovani 1998	Sao Paulo - SP (FES)	34	9	0,273
Borgo et al. 2002	Fenix - PR (FES)	32	10	0,220
Rogalski and Zanin 2003	Marcelino Ramos - RS (FES)	70	8	0,163
Giongo and Waechter 2004	Eldorado do Sul - RS (FES)	57	13	0,317
Breier 2005	Assis - SP (FES)	25	9	0,286
Cervi and Borgo 2007	Foz do Iguaçù - PR (FES)	56	13	0,175
Dettke et al. 2008	Maringa - PR (FES)	29	8	0,232

Keep going...

Table 33 - Continuation...

106

Source	Location (Phytophysiognomy)	Species	Families	Jaccard (J)
Alves and Kolbek 2009	TiradentesZPrados - MG (FES)	61	12	0,165
Menini-Neto *et al.* 2009	Barroso - MG (FES)	41	5	0,081
Menini-Neto *et al.* 2009	Descoberto - MG (FES)	59	10	0,067
Bonnet *et al.* 2011	Tibagi River Basin - PR (FES)	60	13	0,490
Geraldino *et al.* 2010	Campo Mourao - PR (FES/FOM)	61	12	0,247
Lisingen *et al.* 2006	Jaguariaiva - PR (FES/CER)	16	3	0,088

The low similarity between epiphytic richness observed in the central area and that observed in most of the studies shown in Table 33 may be associated with the great sensitivity of the epiphytic community to climatic and microclimatic variations (BENZING, 1990; 1995), given the environmental changes suffered by the forests in the Sorocaba/Médio Tietê basin (ZERO REPORT, 2005), or even the characteristic climate and vegetational peculiarities of each of the studies. Environmental changes, especially the fragmentation and loss of forests, are mainly responsible for the changes and microclimatic variations in forests. Dettke et al. (2008) and Barthlott et al. (2001) point out that the occurrence of epiphytic species is related to the integrity of the forest and consequently to favorable climatic conditions. However, the variations caused by the natural distribution of species in different forest phytophysiognomies and, above all, the occurrence of different species according to latitudinal variations (WAECHTER, 1998), is one possibility for the low similarity between the epiphyte species that occur in the area studied and the species observed by different authors (Table 33). The irregular geographical distribution pointed out by Kersten (2006) may be another important factor behind the low similarity observed above.

It is interesting to note that the greatest similarity occurs with a similar forest formation, or rather, in a similar study carried out by Bonnet et al. (2011) in a river basin in the state of Paranà. Although there is a general tendency for epiphytic diversity to decrease from tropical regions towards the poles (SMITH, 1962), the dependence on humidity, absorbed directly from the air, makes moist forests centers of epiphytic diversity (BENZING, 1990; SCHÜTZ-GATTI, 2000; KERSTEN; SILVA, 2001). This may contribute to the fact that drier forest formations, as in the case of the sites studied, although at lower latitudes, are less diverse than wet forests at higher latitudes.

The quantitative assessment of the vascular epiphytes from all the sites (core and its replicas) in the Central Area recorded the occurrence of 47 species and is shown in Table 34. The Central Area of the Sorocaba/Médio Tiete basin, although characterized as Semideciduous Seasonal Forest, has well-defined climatic characteristics, especially with regard to the distribution of moisture during the year - with colder, drier winters and wetter summers. It is possible to observe a greater importance of species that are resistant to this period of water deficit, such as those belonging to some genera of the Polypodiaceae and Bromeliaceae families, especially Tillandsia and Pleopeltis, and the Cactaceae, which are able to overcome or resist periods of water deficit.

Table 34 - Vascular epiphytes of the Central Area of the Sorocaba/Médio Tietê Hydrographic Basin (Seasonal Semideciduous Forest), classified according to the value of epiphytic importance - nr: absolute number of occurrences in the strata; far: absolute frequency in the strata; ni: absolute number of occurrences in the forophytic individuals; fai: absolute frequency in the forophytic individuals; vt (total value): sum of the abundance estimates; vie: epiphytic importance value; nota: average score obtained.

Species	nr	far	ni	fai	vt	vie	note
Microgramma vacciniifolia	544	25,19	163	45,28	1064	20,73	1,96
Pleopeltis pleopeltifolia	515	23,84	176	48,89	889	17,32	1,73
Tillandsia recurvata	258	11,94	92	25,56	412	8,03	1,60
Tillandsia tricholepis	200	9,26	63	17,50	310	6,04	1,55
Pleopeltis squalida	123	5,69	39	10,83	249	4,85	2,02
Microgramma tecta	131	6,06	39	10,83	240	4,68	1,83
Lepismium cruciforme	105	4,86	38	10,56	213	4,15	2,03
Epiphyllum phyllanthus	96	4,44	48	13,33	164	3,20	1,71
Billbergia distachya	86	3,98	49	13,61	159	3,10	1,85
Aechmea bromeliifolia	76	3,52	43	11,94	156	3,04	2,05
Rhipsalis teres	68	3,15	33	9,17	143	2,79	2,10
Tillandsia stricta	86	3,98	39	10,83	125	2,44	1,45
Lophiaris pumila	58	2,69	34	9,44	112	2,18	1,93
Rhipsalis cereuscula	56	2,59	28	7,78	106	2,07	1,89
Pleopeltis hirsutissima	55	2,55	28	7,78	101	1,97	1,84
Microgramma squamulosa	59	2,73	21	5,83	91	1,77	1,54
Polystachya foliosa	38	1,76	17	4,72	87	1,69	2,29
Microgramma persicariifolia	33	1,53	11	3,06	67	1,31	2,03
Vriesea fenestralis	28	1,30	10	2,78	48	0,94	1,71
Serpocaulon latipes	24	1,11	11	3,06	46	0,90	1,92
Aechmea nudicaulis	19	0,88	12	3,33	43	0,84	2,26
Lepismium warmingianum	19	0,88	8	2,22	34	0,66	1,79
Tillandsia funckiana	19	0,88	10	2,78	28	0,55	1,47
Species	**nr**	**far**	**ni**	**fai**	**vt**	**vie**	**note**
Lepismium lumbricoides	15	0,69	9	2,50	26	0,51	1,73
Peperomia glabella	14	0,65	9	2,50	26	0,51	1,86
Campyloneurum nitidum	12	0,56	7	1,94	21	0,41	1,75
Billbergia porteana	11	0,51	8	2,22	19	0,37	1,73

Philodendron bipinnatifidum	11	0,51	7	1,94	18	0,35	1,64
Cereus alacriportanus	8	0,37	4	1,11	17	0,33	2,13
Anthurium comtum	9	0,42	4	1,11	16	0,31	1,78
Aechmea distichantha	11	0,51	5	1,39	15	0,29	1,36
Epidendrum rigidum	10	0,46	6	1,67	15	0,29	1,50
Peperomia tetraphylla	7	0,32	2	0,56	15	0,29	2,14
Peperomia pereskiifolia	3	0,14	2	0,56	8	0,16	2,67
Pleopeltis astrolepis	4	0,19	3	0,83	7	0,14	1,75
Vittaria lineata	4	0,19	2	0,56	7	0,14	1,75
Campyloneurum centrobrasilianum	3	0,14	3	0,83	6	0,12	2,00
Oeceoclades maculata	3	0,14	2	0,56	5	0,10	1,67
Pecluma filicula	2	0,09	1	0,28	4	0,08	2,00
Peperomia trineuroides	2	0,09	1	0,28	4	0,08	2,00
Tradescantia albiflora	3	0,14	2	0,56	4	0,08	1,33
Asplenium pulchellum	2	0,09	1	0,28	3	0,06	1,50
Rhipsalis pilocarpa	3	0,14	3	0,83	3	0,06	1,00
Aechmea apocalyptica	1	0,05	1	0,28	2	0,04	2,00
Encyclia oncidioides	1	0,05	1	0,28	2	0,04	2,00
Philodendron eximium	1	0,05	1	0,28	2	0,04	2,00
Campyloneurum repens	1	0,05	1	0,28	1	0,02	1,00

The species with the highest value of importance were: *Microgramma vacciniifolia* (Polypodiaceae) with a value of epiphytic importance (VIE) of 20.73 and an average score of 1.96, occurring in 45% of the forophytes and 25% of the strata sampled. *Pleopeltis pleopeltifolia* (Polypodiaceae) with a VIE of 17.32 and an average score of 1.73, occurring in around 49% of the forophytes and 24% of the strata, was the second most important species in the study area. *Tillandsia recurvata* (Bromeliaceae) had a VIE of 8.03 and an average score of 1.6, being found in 25% of the forophytes and 12% of the strata, and *Tillandsia tricholepis* had a VIE of 6.04 and an average score of 1.55. *Pleopeltis squalida* and *Microgramma tecta*, two Polypodiaceae, had a VIE of 4.85 and 4.68 respectively. These six species were responsible for more than 60% of the epiphytic importance value in the Semideciduous Seasonal Forest of the central area in the Sorocaba/Médio Tietê basin. The Polypodiaceae and Bromeliaceae families are among the most frequently observed in Brazilian studies of epiphytes in Semideciduous Seasonal Forest (DISLICH; MANTOVANI, 1998; ROGALSKI; ZANIN, 2003; GIONGO; WAECHTER, 2004; BREIER, 2005; DETTKE et al., 2008). Species from the Cactaceae family are also noteworthy for accounting for almost 14% of the VIE in the central area. Resistance to water deficit and/or temperature variation may be responsible for the success of these genera in this forest formation (BATAGHIN et al., 2012b).

There was no significant difference between the shape of the vertical distribution of the abundances of vascular epiphytes in the Core Site of the Central Area and its three Replicas (p > 0.05). The comparison of the strata of the Core Site and the equivalent strata in Replica Sites I, II and III revealed that the vertical distribution of the vascular epiphytes did not vary significantly in the paired comparisons between the Core Site and its replicas, except for the middle stem and the high stem of Replica Site II, which were significantly different from the equivalent strata in the Core Site.

Looking at the Quantitative Sites in the Central Area, the epiphytic distribution in the strata highlighted the base of the canopy as the stratum with the highest epiphytic abundance (Figure 36), with an abundance value (AV) of 1834, followed by the inner canopy with an AV of 1263, the high stem with an AV of 1024 and the middle stem, outer canopy and low stem with abundance values of 524, 308 and 180 respectively.

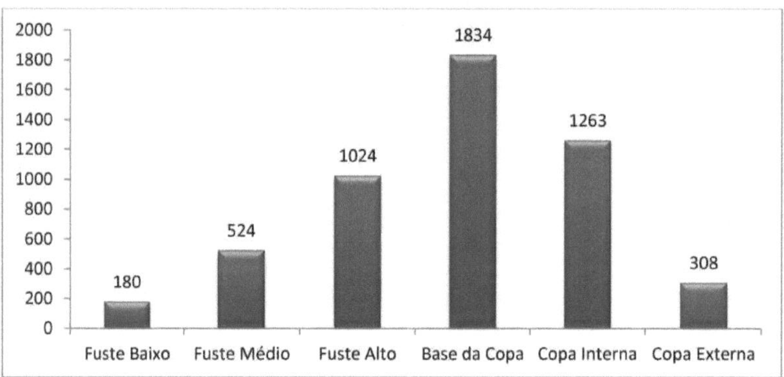

Figura 36:   Distribution of the abundance of epiphytic vascular species among the forophytic strata in the Central Area of the Sorocaba/Médio Tietê river basin.

The analysis applied to the vertical distribution of the total epiphytic community in the Central Area, based on species abundance, showed significant differences between the low stem and all the other strata except the outer canopy (Table 35), indicating the similarity in the number of individuals occurring in these two regions. The medium stem differed from the high stem and the base of the canopy. The outer canopy was significantly different in abundance from the upper canopy, the base of the canopy and the inner canopy. In general, this is indicative of the irregular distribution along the forophytes, with vertical variation in the number of individuals (BROWN, 1990; WAECHTER, 1992). However, the intermediate strata support a greater number of specimens (Figure 36), either because of the favorable microclimate or because of the greater availability of area for installation and development

110

(STEEGE; CORNELISSEN, 1989; ACEBEY; KROMER, 2001).

Table 35: Analysis of the distribution of the richness and abundance of vascular epiphytes among the forophytic strata in the Central Area of the Sorocaba/Médio Tietê basin.

^^^-^^^Richness Abundance^'^^^	Beam Bass	Beam Medium	High stem	World Cup base	Internal Cup	Cup External
Low stem		**0,0188**	0,1136	0,0673	0,5157	**3,39E-05**
Medium stem	**0,003**		0,4272	0,5891	0,0856	**7,11E-10**
High stem	**9,75E-04**	**0,042**		0,8026	0,3492	**3,10E-08**
World Cup base	**3,74E-04**	**0,004**	0,069		0,2348	**9,35E-09**
Internal Cup	**0,008**	0,055	0,322	0,19		**2,22E-06**
External Cup	0,175	0,11	**0,008**	**0,001**	**0,021**	

When analyzing the richness of epiphytic species in the strata, the community that develops in the outer canopy was significantly different from that which occurs in the other strata. This is due to the extreme microclimatic conditions that are characteristic of this part of the forophytes, where there is low humidity and an abundant supply of light. Notably, the number of species capable of growing in this region is small compared to the other parts of the forophyte. In the lower stem, for example, there are 40 of the 47 species recorded in the quantitative survey, while in the outer canopy there are only 10. In general, the species recorded at the upper end of the host trees are not restricted to this region, but occur throughout the forophytes, including in the lower part (low stem). However, species that are more demanding in terms of environmental conditions occur in the lower and middle regions of the forophytes. In addition, it can be seen that the vertical distribution is very similar in all the sites analyzed in the Central Area of the Sorocaba/Médio Tietê basin, which is the result of the so-called vertical evolution of this community (BENZING, 1990) and also of the environmental changes taking place in all the fragments, including the Core Site, which is a Conservation Unit.

The different communities that occur on the forophytes can be seen in Figure 37, which shows the similarity (Jaccard) between the strata.

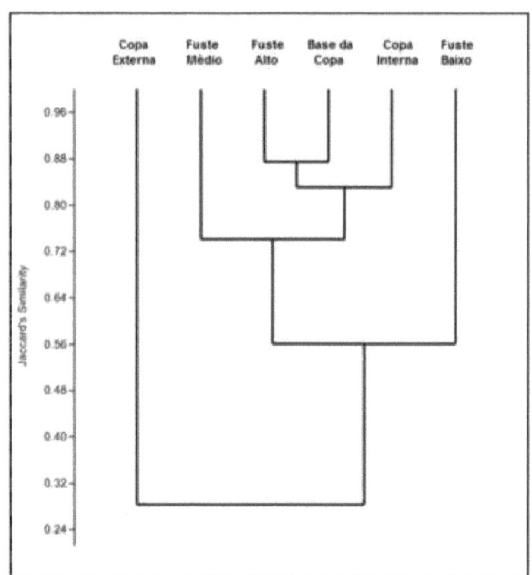

Figura 37:   Dendrogram (UPGMA) of Jaccard's similarity between strata in the Semideciduous Seasonal Forest of the central area of the Sorocaba/Médio Tietê basin.

The greater similarity between the high stem, the base of the canopy and the inner canopy not only reflects the search for a balance between water stress and the availability of light, but also the greater availability of area and better microclimatic conditions for epiphytes. The presence of a more favorable environment for the establishment and survival of epiphytic individuals, since the presence of branches (which are more horizontal) facilitates the anchoring of propagules, especially in anemochoric species, as well as accumulating some amount of soil, retaining nutrients and moisture, favors the development of this community (BENZING, 1990; NIEDER et al., 1999; KERSTEN, 2006). The lower similarity of the outer canopy is related to the presence of a reduced diversity of epiphytes, made up of species capable of withstanding a lack of humidity, excess light and large fluctuations in temperature.

AMOUNT AREA

Floristic survey of vascular epiphytes in the upstream area of the Sorocaba/Médio Tietê river basin

In the floristic survey carried out in the Upstream Area of the Sorocaba/Médio Tietê hydrographic basin, characterized as an ecotone region between the Semideciduous Seasonal Forest and Dense Ombrophilous Forest phytophysiognomies, 139 species were found, belonging to 61 genera and 14 families (Table 36). Shannon's diversity index for the epiphytic vascular community in the upstream area was H'= 3.659, equability J = 0.742 and Margalef's richness index (d) was 16.17.

Table 36 - List of vascular epiphyte species found in the phytosociological survey of the upstream area of the Sorocaba Mèdio Tietê river basin and their respective Ecological Categories (EC) - HLC: Holoepiphyte characteristic; HLF: Holoepiphyte facultative; HLA: Holoepiphyte accidental; HMP: Hemiepiphyte primary and HMS: Hemiepiphyte secondary. Dispersal form (Disp.) - Zo: Zoochoric; An: Anemochoric. Reg.: HUFSCar Herbarium Registration Number (Im = Digital Image).

Family	Species	EC	Disp.	Reg
ARACEAE				
1	Anthurium acutum N.E. Br.	HLA	Zo	8485
2	Anthurium longifolium (Hoffm.) G. Don	HLC	Zo	8499
3	Anthurium sellowianum Kunth	HLF	Zo	Im
4	Philodendron appendiculatum Nadruz & S.J. Mayo	HMP	Zo	8487
5	Philodendron bipinnatifidum Schott	HMP	Zo	Im
6	Philodendron corcovadense Kunth	HMP	Zo	8385
7	Philodendron propinquum Schott	HMS	Zo	8500
8	Philodendron vargealtense Sakur.	HMP	Zo	8394
ASPLENIACEAE				
9	Asplenium auritum Sw.	HLF	An	8497
10	Asplenium mucronatum C. Presl	HLC	An	8483
11	Asplenium pteropus Kaulf.	HLC	An	8481
12	Asplenium scandicinum Kaulf.	HLC	An	8482
13	Asplenium sp.	HLC	An	8438
BEGONIACEAE				
14	Begonia fruticosa A. DC.	HMS	An	8484
BLECHNACEAE				
15	Blechnum binervatum (Poir.) C.V. Morton & Lellinger	HMS	An	8501
BROMELIACEAE				
16	*Acanthostachys Strobilacea* (Schult. f.) Klotzsch	HLC	Zo	8431

Keep going...

Table 36- Continuation...

Family	Species	EC	Disp.	Reg
BROMELIACEAE				
17	*Aechmea bromeliifolia* (Rudge) Baker	HLC	Zo	8432
18	*Aechmea racinae* L.B. Sm.	HLC	Zo	8510
19	*Aechmea distichantha* Lem.	HLC	Zo	8503
20	*Aechmea nudicaulis* (L.) Griseb.	HLC	Zo	8434

21	*Aechmea* sp.	HLC	Zo	8456
22	*Billbergia amoena* (Lodd.) Lindl.	HLC	Zo	8379
23	*Billbergia distachya* (Vell.) Mez	HLC	Zo	8445
24	*Billbergia zebrina* (Herb.) Lindl.	HLC	Zo	8433
25	*Canistropsis billbergioides* (Schult. & Schult.f.) Rudder	HLC	Zo	8511
26	*Canistrum lindenii* (Regel) Mez	HLC	Zo	Im
27	*Neoregelia laevis* (Mez) L.B. Sm.	HLC	Zo	Im
28	*Nidularium rutilans* E. Morren	HLC	Zo	8456
29	*Nidularium innocentii* Lem.	HLC	Zo	8505
30	*Tillandsia araujei* Mez	HLC	An	8377
31	*Tillandsia dura* Baker	HLC	An	8459
32	*Tillandsia fasciculata* Sw.	HLC	An	8514
33	*Tillandsia geminiflora* Brongn.	HLC	An	8427
34	*Tillandsia linearis* Vell.	HLC	An	8462
35	*Tillandsia pohliana* Mez	HLC	An	8463
36	*Tillandsia recurvata* (L.) L.	HLC	An	8429
37	*Tillandsia* sp.	HLC	An	8425
38	*Tillandsia stricta* Sol. ex Sims	HLC	An	8428
39	*Tillandsia tenuifolia* L.	HLC	An	8509
40	*Tillandsia tricholepis* Baker	HLC	An	8430
41	*Tillandsia usneoides* (L.) L.	HLC	An	8460
42	*Vriesea altodaserrae* L.B. Sm.	HLC	An	8461
43	*Vriesea bituminosa* Wawra	HLC	An	8515
44	*Vriesea carinata* Wawra	HLC	An	8513
45	*Vriesea flammea* L.B. Sm.	HLC	An	8507
46	*Vriesea gigantea* Mart. ex Schult. f.	HLC	An	8512
47	*Vriesea hieroglyphica* (Carrière) E. Morren	HLC	An	Im
48	*Vriesea incurvata* Gaudich.	HLC	An	8506
49	*Vriesea platynema* Gaudich.	HLC	An	8464
50	*Vriesea procera* (Mart. ex Schult. & Schult.f.) Wittm.	HLC	An	8508
51	*Vriesea rodigasiana* E. Morren	HLC	An	8504
52	*Vriesea vagans* (L.B. Sm.) L.B. Sm.	HLC	An	Im
CACTACEAE				
53	*Cereus alacriportanus* Pfeiff.	HLF	Zo	Im
54	*Epiphyllum phyllanthus* (L.) Haw.	HLC	Zo	8487
55	*Lepismium cruciforme* (Veil.) Miq.	HLC	Zo	8388

Keep going...

Table 36- Continuation...

Family	Species	EC	Disp.	Reg
CACTACEAE				
56	*Lepismium lumbricoides* (Lem.) Barthlott	HLC	Zo	8386
57	*Rhipsalis baccifera* (J.S. Muell.) Stearn	HLC	Zo	8387
58	*Rhipsalis campos-portoana* Loefgr.	HLC	Zo	8388
59	*Rhipsalis floccosa* Salm-Dyck ex Pfeiff.	HLC	Zo	8380
60	*Rhipsalis paradoxa* (Salm-Dyck ex Pfeiff.) Salm-Dyck	HLC	Zo	8382
61	*Rhipsalis teres* (Vell.) Steud.	HLC	Zo	8384
62	*Rhipsalis trigona* Pfeiff.	HLC	Zo	8381
COMMELINACEAE				
63	*Tradescantia albiflora* Kunth	HLA	Zo	8391
DRYOPTERIDACEAE				
64	*Elaphoglossum lingua* (C. Presl) Brack.	HLC	An	8393
65	*Elaphoglossum glabellum* J.Sm.	HLF	An	8480
66	*Elaphoglossum glaziovii* (Fée) Brade	HLF	An	8479
67	*Elaphoglossum ornatum* (Mett. ex Kuhn) Christ	HLF	An	8476
68	*Polybotrya cylindrica* Kaulf.	HMS	An	8477
69	*Stigmatopteris caudata* (Raddi) C. Chr.	HLA	An	8478
GESNERIACEAE				
70	*Codonanthe devosiana* Lem.	HLC	Zo	Im
71	*Codonanthe gracilis* (Mart.) Hanst.	HLC	Zo	8491
72	*Nematanthus striatus* (Handro) Chautems	HLC	Zo	8489
73	*Sinningia douglasii* (Lindl.) Chautems	HLC	Zo	8490
MARCGRAVIACEAE				
74	*Marcgravia polyantha* Delpino	HMS	Zo	8502
ORCHIDACEAE				
75	*Anathallis sclerophylla* (Lindl.) Pridgeon & M.W. Chase	HLC	An	8515
76	*Brasilidium* sp.	HLC	An	8398
77	*Brasiliorchis gracilis* (Lindl.) R.B. Singer *et al.*	HLC	An	8396
78	*Bulbophyllum napellii* Lindl.	HLC	An	8414
79	*Campylocentrum aromaticum* Barb.Rodr.	HLC	An	8419
80	*Campylocentrum* cf. *grisebachii* Cogn.	HLC	An	Im
81	*Capanemia micromera* Barb. Rodr.	HLC	An	8423
82	*Catasetum fimbriatum* (C.Morren) Lindl.	HLC	An	8424
83	*Catasetum atratum* Lindl.	HLC	An	8451
84	*Catasetum* sp.	HLC	An	8496
85	*Cattleya* sp.	HLC	An	Im
86	*Coppensia varicosa* (Lindl.)Campacci	HLC	An	8406

87	*Cyclopogon multiflorus* Schltr.	HLA	An	8404
88	*Dichaea pendula* (Aubl.) Cogn.	HLC	An	8495
89	*Dichaea trulla* Rchb. f.	HLC	An	8470
90	*Encyclia patens* Hook.	HLC	An	8493

Keep going...

Table 36- Continuation...

Family	Species	EC	Disp.	Reg
ORCHIDACEAE				
91	*Epidendrum ansiferum* Rchb. f.	HLc	An	8373
92	*Gomesa recurva* R. Br.	HLc	An	8410
93	*Gomesa glaziovii* cogn.	HLc	An	8411
94	*Gomesa* sp.	HLc	An	8417
95	*Grobya galeata* Lindl.	HLc	An	8405
96	*Lophiaris pumila* (Lindl.) Braem	HLc	An	8468
97	*Miltonia* sp.	HLc	An	Im
98	*Notylia longispicata* Hoehne & Schltr.	HLc	An	8416
99	*Octomeria grandiflora* Lindl.	HLc	An	8418
100	*Polystachya estrellensis* Rchb.f.	HLc	An	8467
101	*Polystachya concreta* (Jacq.) Garay & H.R. Sweet	HLc	An	8496
102	*Prosthechea glumacea* (Lindl.) W.E. Higgins	HLc	An	8399
103	*Prosthechea* cf. *bulbosa* (Vell.) W.E. Higgins	HLc	An	Im
104	*Rodriguezia decora* (Lem.) Rchb.f.	HLc	An	8441
105	*Saundersia mirabilis* Rchb.f.	HLc	An	8494
106	*Scaphyglottis modesta* (Rchb. f.) Schltr.	HLc	An	8378
107	*Stelis deregularis* Barb. Rodr.	HLc	An	8473
PIPERAcEAE				
108	*Peperomia alata* Ruiz & Pav.	HLc	Zo	8420
109	*Peperomia castelosensis* Yunck.	HLc	Zo	8421
110	*Peperomia catharinae* Miq.	HLc	Zo	8422
111	*Peperomia glabella* (Sw.) A. Dietr.	HLF	Zo	8475
112	*Peperomia pereskiifolia* (Jacq.) Kunth	HLc	Zo	8448
113	*Peperomia trineura* Miq.	HLc	Zo	8446
114	*Peperomia urocarpa* Fisch. & c.A. Mey.	HLc	Zo	8492
POLYPODIAcEAE				
115	*Campyloneurum acrocarpon* Fée	HLc	An	8529
116	*Campyloneurum nitidum* (Kaulf.) c. Presl	HLc	An	8531
117	*Campyloneurum repens* (Aubl.) c. Presl	HLc	An	8533
118	*Ceradenia albidula* (Baker) L.E. Bishop	HLc	An	8376

119	*Microgramma tecta* (Kaulf.) Alston	HLc	An	8524
120	*Microgramma crispata* (Fée) R.M.Tryon & A.F.Tryon	HLc	An	8530
121	*Microgramma lycopodioides* (L.) copel.	HLc	An	8437
122	*Microgramma percussa* (cav.) de la Sota	HLc	An	8392
123	*Microgramma persicariifolia* (Schrad.) c. Presl	HLc	An	8520
124	*Microgramma squamulosa* (Kaulf.) de la Sota	HLc	An	8534
125	*Microgramma vacciniifolia* (Langsd. & Fisch.) copel.	HLc	An	8522
126	*Niphidium crassifolium* (L.) Lellinger	HLc	An	8516
127	*Pecluma filicula* (Kaulf.) M.G. Price	HLc	An	8527
128	*Pecluma* sp.	HLC	An	8375

Keep going...

Table 36- Continuation...

Family	Species	EC	Disp.	Reg
POLYPODIACEAE				
129	*Pecluma truncorum* (Lindm.) M.G. Price	HLC	An	8535
130	*Pleopeltis hirsutissima* (Raddi) de la Sota	HLC	An	8521
131	*Pleopeltis macrocarpa* (Bory ex Willd.) Kaulf.	HLC	An	8435
132	*Pleopeltis pleopeltifolia* (Raddi) Alston	HLC	An	8525
133	*Pleopeltis squalida* (Vell.) de la Sota	HLC	An	8436
134	*Serpocaulon catharinae* (Langsd. & Fisch.) A.R. Sm.	HLC	An	8528
135	*Serpocaulon fraxinifolium* (Jacq.) A.R. Sm.	HLC	An	8518
136	*Serpocaulon latipes* (Langsd. & Fisch.) A.R. Sm	HLC	An	8423
137	*Serpocaulon sehnemii* (Pic.-Serm.) Labiak & J.Prado	HLC	An	8532
PTERIDACEAE				
138	*Polytaenium cajenense* (Desv.) Benedict	HLC	An	8439
139	*Vittaria lineata* (L.) Sm.	HLC	An	8454

The specific richness of the area can be considered low when compared to other studies carried out in dense ombrophilous forest or wet forest formations. In Dense Ombrophilous Forest, authors report a higher number of epiphyte species, e.g. Blum (2010) - 277 species; Kersten (2006) - 349 species; Fontoura et al. (1997) - 293 species. However, the numbers obtained here are higher or similar to those obtained by Hertel (1950) - 101 species; Petean (2003) - 97 species; Menini-Neto et al. (2009) - 113 species; Schütz-Gatti (2000) - 175 species; Breier (2005) - 161 species; and Petean (2009) - 159 species. The record of a considerable number of epiphytic species in the Upstream Area of the Sorocaba Médio Tietê basin, in relation to the other collection areas, reinforces the idea of dependence on atmospheric humidity (GENTRY; DODSON, 1987a), since the acquisition and storage of water are the most relevant factors for epiphytic growth (ZOTS; HIETZ, 2001).

The richness of epiphytic species found in the area is greater than that observed in the surveys in Semideciduous Seasonal Forest carried out by Rogalski and Zanin (2003) who found 70 species, by Giongo and Waechter (2004) who sampled 57 species, by Cervi and Borgo (2007) who found 56 species, by Aguiar et al. (1981), who sampled 17 species, by Dislich and Mantovani (1998), with 34 species, by Borgo et al. (2002), with 32 species, by Breier (2005), with 25 species and by Dettke et al. (2008), with 29 species.

The presence of epiphytic species in ecotone areas between two forest formations is generally greater than both adjacent communities (BONNET et al., 2011). However, competition in the ecotone environment between the different forests can generate a new community composition, with species shared by both phytophysiognomies, but which is not more diverse than the phytophysiognomy with the highest diversity. Many epiphytic species can be excluded, not only because of the specific competition that exists, but also because of the abiotic factors (greater luminosity and lower water supply) that affect the ecotone differently from the adjacent plant formations.

In addition, Blum (2010), studying epiphytes on different altitudinal gradients in Morretes - PR, revealed the occurrence of 121 epiphytic species in the altitudinal range corresponding to the Sub-Montane and Montane Dense Ombrophylous Forest. The sites studied in the Montante Area are located in a forest formation (vegetation type) similar to that found in the study by Blum (2010), noting that 139 species of vascular epiphytes were identified in the Montante Area.

In the survey of the Ecotone between Semideciduous Seasonal Forest and Dense Ombrophilous Forest in the Montante Area of the Sorocaba/Médio Tietê river basin, the epiphytic families with the highest species richness were: Bromeliaceae (37 species), Orchidaceae (33 species), Polypodiaceae (23 species), Cactaceae (10 species) Araceae (eight species), Piperaceae (seven species), Dryopteridaceae (six species), Aspleniaceae (five species), Gesneriaceae (four species) and Pteridaceae (two species). The families Begoniaceae, Blechnaceae, Commelinaceae and Marcgraviaceae had only one species. The distribution of epiphytic species in the ecological categories (Figure 38), according to the relationship with the forophyte proposed by Benzing (1990), showed a predominance of characteristic holoepiphytes with 118 species (84%), followed by facultative holoepiphytes with eight species (6%), secondary hemiepiphytes with five species (4%), accidental holoepiphytes with four species (3%), and primary hemiepiphytes also with four species (3%). The predominance of characteristic holoepiphytes has been observed in Dense Ombrophilous Forest in several studies (BLUM, 2010; PETEAN, 2009; KERSTEN, 2006;

BREIER, 2005; FONTOURA et al., 1997; SCHÜTZ-GATTI, 2000; PETEAN, 2003) and in Semideciduous Seasonal Forest (PINTO et al., 1995; DISLICH; MANTOVANI, 1998; ROGALSKI; ZANIN, 2003; CERVI; BORGO, 2007; DETTKE et al., 2008; BATAGHIN et al., 2010), and in other forest formations, such as Mixed Ombrophilous Forest (DITTRICH et al., 1999), in Cerrado areas (BREIER 2005; BATAGHIN et al., 2012b) and in restinga areas (WAECHTER, 1992; KERSTEN; SILVA, 2001).

Dispersal strategy is an important factor in the success of epiphytic synopsis (GENTRY; DODSON, 1987a), and anemochory predominates as a dispersal syndrome among epiphytic species (BENZING, 1987; BREIER, 2005; DETTKE et al., 2008; MENINI-NETO et al., 2009; GERALDINO et al., 2010). The results of this study corroborate this idea, as 68% of the epiphytic species showed anemochoric dispersal, while only 32% showed zoochory as a dispersal syndrome. This high percentage of anemochory is a reflection of the large number of orchids, ferns and bromeliads (in the latter case, especially the genera *Tillandsia* and *Vriesea)* recorded in the study area.

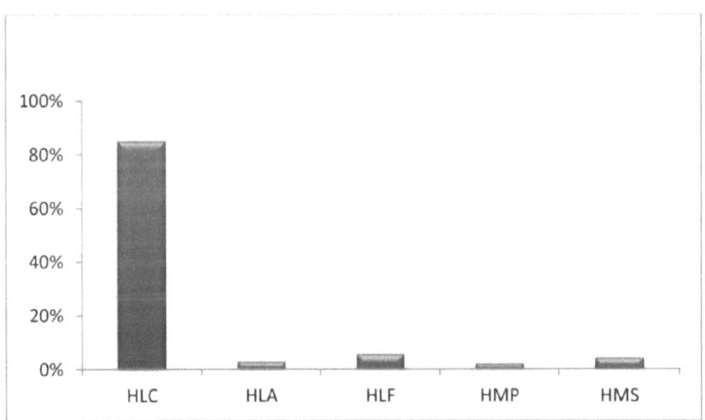

Figura 38:   Distribution of the epiphytic vascular species of the Semideciduous Seasonal Forest and Dense Ombrophilous Forest ecotone in the upstream area of the Sorocaba/Médio Tietê hydrographic basin in the ecological categories proposed by Benzing (1990): HLC: characteristic holoepiphytes; HLF: facultative holoepiphytes; HLA: accidental holoepiphytes; HMP: primary hemiepiphytes; HMS: secondary hemiepiphytes.

The Bromeliaceae, Orchidaceae and Polypodiaceae families are responsible for 67% (93 species) of the species found in the floristic survey, a percentage very similar to that found by Kersten (2006), Petean (2009) and Blum (2010). Furthermore, these families are considered to be the richest in epiphytes worldwide (MADISON, 1977; KRESS, 1986; GENTRY; DODSON, 1987b; BENZING, 1990). The Araceae and Cactaceae families, which had eight

(6%) and 10 (7%) species respectively in the study area, are also worth mentioning. Araceae is usually more prominent in Dense Ombrophilous Forest than in drier forests (KERSTEN, 2006), however, the works by Petean (2009) and Blum (2010), carried out in Dense Ombrophilous Forest, show a similar number of species for these families. Cactaceae also deserves to be highlighted, as despite accounting for only 0.5% of the world's epiphytic species (MADISON, 1977; BENZING, 1990) and 3% of Brazilian epiphytes (KERSTEN, 2006), in the study area they presented 10 species each, making up more than 7% of the total epiphytic species in the study area.

It is important to note that the Orchidaceae family, which is the richest epiphyte family worldwide (MADISON, 1977; KRESS, 1986; BENZING, 1990), in the Neotropics (GENTRY; DODSON, 1987a) and in Brazil (KERSTEN, 2006), had 33 species, a lower number than those found by Kersten (2006) - 47 species (FOD/FOM), Petean (2009) - 61 species (FOD) and Blum (2010) - 103 species (FOD). However, Blum (2010) found 32 orchid species in Dense Ombrophilous Forest, in the same altitudinal range as the research presented here, and these numbers are similar to those presented here.

The number of orchids is also similar to that observed by Rogalski and Zanin (2003), who found 38 species in the Semideciduous Seasonal Forest (FES), and higher than the results obtained by Giongo and Waechter (2004) - 16 species (FES), by Cervi and Borgo (2007) - nine species (FES) and by Bonnet et al. (2011) - 16 species (FES), and much higher than the number published by Dislich and Mantovani (1998) - 6 species (FES), Breier (2005) - 3 species (FES), Dettke et al. (2008) - 3 species (FES), Bataghin et al. (2010) - 2 species (FES), Breier (2005) - 3 species (Cerrado) and Bataghin et al. (2012b) - 8 species (Cerrado). Stancato et al. (2002) suggest that high light intensity can reduce the growth and development of orchids. It seems that the climatic variations characteristic of the study area have some influence on these numbers, so that areas with higher humidity tend to have a greater diversity of epiphytic orchids.

Despite the difficult access to the study areas (most of which are steep), it is important to report visible signs of predatory collection of species in the Montante Area, especially orchids; among these signs, we can mention the presence of trails, traces of climbing and even the cutting down of trees to remove epiphytes, a fact that may be related to the relatively low representation of the Orchidaceae family. Ditt (2002) highlights anthropogenic interference, microclimatic changes resulting from environmental modification and the possible illegal collection of specimens (especially those of commercial - ornamental - interest) as factors that can lead to a reduction in the number of individuals, contributing to a decrease in biological diversity and environmental degradation.

Distribution of vascular epiphytes in the sample sites in the Upland Area

Analysis of the vascular epiphytes of the Core Site of the Semideciduous Seasonal Forest/Dense Ombrophilous Forest Ecotone in the Upstream Area of the Sorocaba/Médio Tietê basin

The survey of epiphytes carried out in the Core Site of the Upstream Area of the Sorocaba/Médio Tietê Basin 1 took place in the northern glebe of the Jurupara State Park (PEJU), a conservation unit located between coordinates 23°50' to 24°02' S and 47°13' to 47°23' W, which covers 26.250.47 ha in the municipalities of Ibiùna and Piedade in the state of São Paulo (sampling site UTM 267.066 and 7.363.092 of zone 23 south). The Conservation Unit was established by Decree No. 35.703, of 22/09/92, which established the Park, with an area of 23,900.47 ha, and Decree No. 35.704, of 22/09/92, which incorporated an additional area of 2,350 ha donated by CBA into the PEJU's boundaries, giving it its current configuration and area. The relief of PEJU (Figure 39) has slopes ranging from 5% to 20% and the region's climate is characterized as Cfb, humid subtropical, with no defined dry season (generally drier winters), with an average annual temperature of 22 °C (SÂO PAULO, 2010).

Figura 39:   Aspect of the relief and vegetation of Juruparà State Park, Sitio Core da Area Montante.

The importance of the PEJU Conservation Unit for biodiversity conservation is due not only to the extent of the area covered by different phytophysiognomies, but also to the complex microclimatic gradient that favors the existence and establishment of animal and plant species. To date, 1,144 species belonging to 230 families have been recorded. The flora of this Conservation Unit includes 557 species, however, the Management Plan for this

121

Conservation Unit indicates that the richness of plant groups is unknown, among which the epiphytes stand out (SÃO PAULO, 2010).

In the floristic survey of Sitio Core, 80 species were recorded, belonging to 41 genera and 13 families (Table 37). The Shannon index for the site was H' = 3.456, the equability (J) was 0.831 and the Margalef richness (d) was 8.340.

Table 37 - List of vascular epiphyte species found in the Core Site of the Montante Area (Jurupará State Park) of the Sorocaba/Médio Tietê river basin and their respective Ecological Categories (EC) - HLC: Holoepiphyte characteristic; HLF: Holoepiphyte facultative; HLA: Holoepiphyte accidental; HMP: Hemiepiphyte primary: HMS: Hemiepiphyte secondary. Dispersal form (Disp.) - Zo: Zoochoric; An: Anemochoric. Reg: HUFSCar herbarium registration number (Im: Digital image).

Family	Species	EC	Disp.	Reg.
ARACEAE				
1	Anthurium acutum N.E. Br.	HLA	Zo	8485
2	Anthurium longifolium (Hoffm.) G. Don	HLC	Zo	8499
3	Anthurium sellowianum Kunth	HLF	Zo	Im
4	Philodendron appendiculatum Nadruz & S.J. Mayo	HMP	Zo	8487
5	Philodendron corcovadense Kunth	HMP	Zo	8385
6	Philodendron propinquum Schott	HMS	Zo	8500
7	Philodendron vargealtense Sakur.	HMP	Zo	8394
ASPLENIACEAE				
8	Asplenium auritum Sw.	HLF	An	8497
9	Asplenium mucronatum C. Presl	HLC	An	8483
10	Asplenium scandicinum Kaulf.	HLC	An	8482
11	Asplenium sp.	HLC	An	8438
BEGONIACEAE				
12	Begonia fruticosa (Klotzsch) A. DC.	HMS	An	8484
BLECHNACEAE				
13	Blechnum binervatum (Poir.) C.V. Morton & Lellinger	HMS	An	8501
BROMELIACEAE				
14	Aechmea racinae L.B. Sm.	HLC	Zo	8510
15	Aechmea nudicaulis (L.) Griseb.	HLC	Zo	8434
16	*Canistropsis billbergioides* (Schult. & Schult.f.) Rudder	HLC	Zo	8511

Table 37- Continuation...

Family	Species	EC	Disp.	Reg.
BROMELIACEAE				

17	*Canistrum lindenii* (Regel) Mez	HLC	Zo	Im
18	*Neoregelia laevis* (Mez) L.B. Sm.	HLC	Zo	Im
19	*Nidularium rutilans* E. Morren	HLC	Zo	8456
20	*Nidularium innocentii* Lem.	HLC	Zo	8505
21	*Tillandsia dura* Baker	HLC	An	8459
22	*Tillandsia geminiflora* Brongn.	HLC	An	8427
23	*Tillandsia linearis* Vell.	HLC	An	8462
24	*Tillandsia pohliana* Mez	HLC	An	8463
25	*Tillandsia* sp.	HLC	An	8425
26	*Tillandsia usneoides* (L.) L.	HLC	An	8460
27	*Vriesea altodaserrae* L.B. Sm.	HLC	An	8461
28	*Vriesea bituminosa* Wawra	HLC	An	8515
29	*Vriesea carinata* Wawra	HLC	An	8513
30	*Vriesea flammea* L.B. Sm.	HLC	An	8507
31	*Vriesea gigantea* Mart. ex Schult. f.	HLC	An	8512
32	*Vriesea hieroglyphica* (Carrière) E. Morren	HLC	An	Im
33	*Vriesea incurvata* Gaudich.	HLC	An	8506
34	*Vriesea platynema* Gaudich.	HLC	An	8464
35	*Vriesea rodigasiana* E. Morren	HLC	An	8504
36	*Vriesea vagans* (L.B. Sm.) L.B. Sm.	HLC	An	Im

CACTACEAE

37	*Lepismium lumbricoides* (Lem.) Barthlott	HLC	Zo	Im
38	*Rhipsalis campos-portoana* Loefgr.	HLC	Zo	8389
39	*Rhipsalis paradoxa* (Salm-Dyck ex Pfeiff.) Salm-Dyck	HLC	Zo	8382
40	*Rhipsalis teres* (Vell.) Steud.	HLC	Zo	8384

COMMELINACEAE

| 41 | *Tradescantia albiflora* Kunth | HLA | Zo | 8391 |

DRYOPTERIDACEAE

42	*Elaphoglossum glabellum* J.Sm.	HLF	An	8480
43	*Elaphoglossum glaziovii* (Fée) Brade	HLF	An	8479
44	*Elaphoglossum lingua* (C. Presl) Brack.	HLC	An	8393
45	*Elaphoglossum ornatum* (Mett. ex Kuhn) Christ	HLF	An	8476
46	*Stigmatopteris caudata* (Raddi) C. Chr.	HLA	An	8478

GESNERIACEAE

47	*Codonanthe devosiana* Lem.	HLC	Zo	Im
48	*Codonanthe gracilis* (Mart.) Hanst.	HLC	Zo	8491
49	*Nematanthus striatus* (Handro) Chautems	HLC	Zo	8489
50	*Sinningia douglasii* (Lindl.) Chautems	HLC	Zo	8490

MARCGRAVIACEAE

| 51 | *Marcgravia polyantha* Delpino | HMS | Zo | 8502 |

Keep going...

Table 37- Continuation...

Family	Species	EC	Disp.	Reg.
ORCHIDACEAE				
52	*Anathallis sclerophylla* (Lindl.) Pridgeon & M.W. chase	HLc	An	8515
53	*Brasilidium* sp.	HLc	An	8398
54	*Catasetum atratum* Lindl.	HLc	An	8451
55	*Catasetum fimbriatum* (c.Morren) Lindl.	HLc	An	8424
56	*Dichaea pendula* (Aubl.) cogn.	HLc	An	8495
57	*Dichaea trulla* Rchb. f.	HLc	An	8470
58	*Encyclia patens* Hook.	HLc	An	8493
59	*Epidendrum ansiferum* Rchb. f.	HLc	An	8373
60	*Gomesa recurva* R. Br.	HLc	An	8410
61	*Grobya galeata* Lindl.	HLc	An	8405
62	*Octomeria grandiflora* Lindl.	HLc	An	8418
63	*Polystachya concreta* (Jacq.) Garay & H.R. Sweet	HLc	An	8496
64	*Prosthechea* cf. *bulbosa* (Vell.) W.E. Higgins	HLc	An	Im
65	*Prosthechea glumacea* (Lindl.) W.E. Higgins	HLc	An	8399
66	*Scaphyglottis modesta* (Rchb. f.) Schltr.	HLc	An	8378
67	*Stelis deregularis* Barb. Rodr.	HLc	An	8473
POLYPODIAcEAE				
68	*Campyloneurum acrocarpon* Fée	HLc	An	8529
69	*Campyloneurum repens* (Aubl.) c. Presl	HLc	An	8533
70	*Microgramma lycopodioides* (L.) copel.	HLc	An	8437
71	*Microgramma persicariifolia* (Schrad.) c. Presl	HLc	An	8520
72	*Microgramma vacciniifolia* (Langsd. & Fisch.) copel.	HLc	An	8522
73	*Niphidium crassifolium* (L.) Lellinger	HLc	An	8516
74	*Pecluma truncorum* (Lindm.) M.G. Price	HLc	An	8535
75	*Pleopeltis hirsutissima* (Raddi) de la Sota	HLc	An	8521
76	*Pleopeltis macrocarpa* (Bory ex Willd.) Kaulf.	HLc	An	8435
77	*Serpocaulon catharinae* (Langsd. & Fisch.) A.R. Sm.	HLc	An	8528
78	*Serpocaulon fraxinifolium* (Jacq.) A.R. Sm.	HLc	An	8518
79	*Serpocaulon latipes* (Langsd. & L. Fisch.) A.R. Sm.	HLc	An	8423
PTERIDAcEAE				
80	*Vittaria lineata* (L.) Sm.	HLC	An	8454

The richness of epiphytic species found in the northern glebe of Juruparà State Park (Sitio Core - Area Montante) can be considered low, especially when compared to studies carried out in Dense Ombrophilous Forest by Blum (2010) - 277 species, Petean (2009) - 159 species, Kersten (2006) - 349 species, Breier (2005) - 161 species, Fontoura et al. (1997) - 293 species, and Schütz-Gatti (2000) - 175 species; however, the numbers are similar to those obtained by Hertel (1950) - 101 species and Petean (2003) - 97 species. When compared to studies carried out in Semideciduous Seasonal Forest, the numbers are higher than those of Rogalski and Zanin (2003) who found 70 species, Bonnet et al. (2011), who observed 60 species, Giongo and Waechter (2004) who sampled 57 species, Cervi and Borgo (2007) who found 56 species, Aguiar et al. (1981), who sampled 17 species, Dislich and Mantovani (1998), with 34 species, Borgo et al. (2002), with 32 species, Breier (2005), with 25 species, and Dettke et al. (2008), with 29 species.

The vascular epiphytic richness of Sitio Core may have suffered or still be suffering from anthropic interference, given that there are around 80 families living within the Conservation Unit, and the great external access of people. In addition, the Juruparà State Park management plan itself identifies activities/actions that could jeopardize the conservation of biodiversity in the UC area, including hunting, fishing, palm heart extraction, the presence of domestic animals, fish farming (exotic/translocated), and the presence of structures related to energy generation and transmission (SÃO PAULO, 2010). Several authors have reported the loss of epiphytic diversity as a result of human interference in environments (BARTHLOTT et al., 2001; WOLF, 2005; BATAGHIN et al., 2008; DETTKE et al., 2008).

At Sitio Core, Bromeliaceae was the richest family with 23 species. Orchidaceae had 16 species, Polypodiaceae had 12, Araceae, seven, Dryopteridaceae, five, and the families Aspleniaceae, Cactaceae and Gesneriaceae had four species each. The families Begoniaceae, Blechnaceae, Commelinaceae, Marcgraviaceae and Pteridaceae had only one species each. Characteristic holoepiphytes were dominant with 81% of the species, followed by facultative holoepiphytes (6%), secondary hemiepiphytes (5%), primary hemiepiphytes and accidental holoepiphytes, both categories with 4% of the species in the Core Site of the Upland Area. This predominance of characteristic holoepiphytes has often been observed in dense ombrophilous forests (BLUM, 2010; PETEAN, 2009; KERSTEN, 2006; BREIER, 2005; FONTOURA et al, 1997; SCHÜTZ-GATTI, 2000; PETEAN, 2003) and also in Semideciduous Seasonal Forest (PINTO et al., 1995; DISLICH; MANTOVANI, 1998; ROGALSKI; ZANIN, 2003; CERVI; BORGO, 2007; DETTKE et al., 2008; BATAGHIN et al., 2010),

As for the dispersal syndrome, 56 epiphytic species (70%) were anemochoric and 24 species (30%) were zoochoric. This pattern of two thirds of epiphytic species being anemochoric

(BENZING, 1987) has been observed in several studies in Brazil (BREIER, 2005; DETTKE et al., 2008; MENINI-NETO et al., 2009; GERALDINO et al. 2010) and was also observed for the Upstream Area of the watershed.

The quantitative analysis recorded 64 species and highlighted a Bromeliaceae as the most important species at Sitio Core (Table 38): *Vriesea incurvata* (Figure 40) was responsible for almost 18% of the epiphytic importance value (EVI) recorded at this site. The species had a VIE of 17.92 and an average score of 1.65, and was recorded in almost 75% of the forophytes and almost 40% of the strata. This contributed to the Bromeliaceae family, with 37.21% of the EVI, being highlighted as the most important in the Core Site.

Table 38 - Vascular epiphytes of the Core Site of the Montante Area (Juruparâ State Park - northern glebe) of the Sorocaba/Médio Tietê hydrographic basin, classified according to the value of epiphytic importance - nr: absolute number of occurrences in the strata; far: absolute frequency in the strata; ni: absolute number of occurrences in the forophytic individuals; fai: absolute frequency in the forophytic individuals; vt (total value): sum of the abundance estimates; vie: epiphytic importance value; nota: average score obtained.

Species	nr	far	ni	fai	vt	vie	note
Vriesea incurvata	207	38.3	67	74.4	342	17,92	1.65
Dichaea trulla	92	17.0	40	44.4	146	7,65	1.59
Philodendron propinquum	67	12.4	23	25.6	109	5,71	1.63
Vriesea altodaserrae	43	8.0	20	22.2	81	4,25	1.88
Pecluma truncorum	43	8.0	17	18.9	69	3,62	1.60
Vriesea carinata	46	8.5	18	20.0	66	3,46	1.43
Serpocaulon catharinae	43	8.0	21	23.3	65	3,41	1.51
Anthurium sellowianum	29	5.4	7	7.8	61	3,20	2.10
Sinningia douglasii	35	6.5	13	14.4	58	3,04	1.66
Begonia fruticosa	35	6.5	12	13.3	53	2,78	1.51
Tillandsia geminiflora	41	7.6	14	15.6	53	2,78	1.29
Campyloneurum acrocarpon	25	4.6	12	13.3	45	2,36	1.80
Codonanthe gracilis	20	3.7	5	5.6	42	2,20	2.10
Marcgravia polyantha	25	4.6	9	10.0	42	2,20	1.68
Serpocaulon latipes	27	5.0	12	13.3	42	2,20	1.56
Campyloneurum repens	23	4.3	11	12.2	39	2,04	1.70
Vriesea gigantea	16	3.0	11	12.2	37	1,94	2.31
Neoregelia laevis	16	3.0	9	10.0	36	1,89	2.25
Vriesea rodigasiana	22	4.1	11	12.2	33	1,73	1.50
Elaphoglossum glabellum	17	3.1	8	8.9	30	1,57	1.76

Table 38 - Continuation...

Species	nr	far	ni	fai	vt	vie	note
Microgramma persicariifolia	16	3.0	5	5.6	30	1,57	1.88
Serpocaulon fraxinifolium	17	3.1	7	7.8	29	1,52	1.71
Microgramma lycopodioides	16	3.0	6	6.7	28	1,47	1.75
Vittaria lineata	22	4.1	10	11.1	28	1,47	1.27
Elaphoglossum ornatum	19	3.5	9	10.0	24	1,26	1.26
Asplenium scandicinum	15	2.8	8	8.9	23	1,21	1.53
Niphidium crassifolium	13	2.4	7	7.8	22	1,15	1.69
Canistrum lindenii	10	1.9	8	8.9	21	1,10	2.10
Asplenium mucronatum	12	2.2	4	4.4	19	1,00	1.58
Anthurium acutum	11	2.0	6	6.7	15	0,79	1.36
Pleopeltis hirsutissima	9	1.7	6	6.7	15	0,79	1.67
Vriesea hieroglyphica	7	1.3	4	4.4	15	0,79	2.14
Asplenium sp.	8	1.5	3	3.3	14	0,73	1.75
Scaphyglottis modesta	7	1.3	4	4.4	13	0,68	1.86
Brasilidium sp.	7	1.3	5	5.6	12	0,63	1.71
Elaphoglossum lingua	6	1.1	5	5.6	11	0,58	1.83
Epidendrum ansiferum	5	0.9	3	3.3	10	0,52	2.00
Philodendron vargealtense	5	0.9	2	2.2	10	0,52	2.00
Tillandsia dura	7	1.3	4	4.4	10	0,52	1.43
Anthurium longifolium	5	0.9	4	4.4	9	0,47	1.80
Dichaea pendula	7	1.3	4	4.4	9	0,47	1.29
Philodendron appendiculatum	4	0.7	2	2.2	7	0,37	1.75
Rhipsalis teres	3	0.6	2	2.2	7	0,37	2.33
Vriesea platynema	4	0.7	3	3.3	7	0,37	1.75
Asplenium auritum	4	0.7	2	2.2	6	0,31	1.50
Grobya galeata	3	0.6	2	2.2	6	0,31	2.00
Lepismium lumbricoides	2	0.4	1	1.1	5	0,26	2.50
Nidularium rutilans	2	0.4	1	1.1	5	0,26	2.50
Prosthechea glumacea	3	0.6	2	2.2	5	0,26	1.67
Catasetum atratum	2	0.4	2	2.2	4	0,21	2.00
Catasetum fimbriatum	2	0.4	2	2.2	4	0,21	2.00
Octomeria grandiflora	2	0.4	2	2.2	4	0,21	2.00
Philodendron corcovadense	3	0.6	1	1.1	4	0,21	1.33
Tillandsia sp.	3	0.6	2	2.2	4	0,21	1.33
Blechnum binervatum	2	0.4	1	1.1	3	0,16	1.50
Elaphoglossum glaziovii	2	0.4	1	1.1	3	0,16	1.50

Encyclia patens	2	0.4	1	1.1	3	0,16	1.50
Prosthechea cf. *bulbosa*	2	0.4	1	1.1	3	0,16	1.50
Gomesa recurva	1	0.2	1	1.1	2	0,10	2.00
Microgramma vacciniifolia	2	0.4	1	1.1	2	0,10	1.00
Nematanthus striatus	1	0.2	1	1.1	2	0,10	2.00
Polystachya concreta	1	0.2	1	1.1	2	0,10	2.00
Rhipsalis campos-portoana	1	0.2	1	1.1	2	0,10	2.00
Stigmatopteris caudata	1	0.2	1	1.1	2	0,10	2.00

The Polypodiaceae family, although without any significant VIE species, was the second most important family in the core site, with a total VIE of 20.23. The Orchidaceae family was the third most important in the site (VIE = 11.69), a result influenced by the recording of the species Dichaea trulla, which obtained a VIE = 7.65 and an average score of 1.59. The Araceae family had the fourth highest importance value - 11.27, and Philodendron propinquum with a VIE = 5.71 and an average score of 1.63 was the main species in this family, accounting for more than 50% of the importance value. Gesneriaceae also deserves a mention, because although it only had four species, it had a VIE = 7.55.

Figura 40:   Vriesea incurvata Gaudich. (Bromeliaceae), the most important epiphytic species in the Core Site of the Upland Area.

The representativeness of the Bromeliaceae, Polypodiaceae and Orchidaceae families may be related to the fact that they are among the richest epliphyte families in the world (GENTRY; DODSON, 1987b), and in Brazil (KERSTEN, 2006). As for the Araceae family, Kersten (2006) highlights the occurrence of this family in moist forests, especially in the Dense

128

Ombrophilous Forest. The same author also points out that around 11% of epiphytic species in areas of Dense Ombrophilous Forest belong to the Araceae family. Also noteworthy is the Gesneriaceae family, which tended to be recorded only in areas of dense ombrophilous forest.

The richness of epiphytic species, although lower than that recorded in other studies in the same type of forest, cannot be considered low for the study area in question, especially if we look at the richness data for areas of seasonal forest surrounding the core site. Although the occurrence of epiphytic species is influenced by the condition of the forest, especially if it suffers from anthropogenic interference (BARTHLOTT et al., 2001; BATAGHIN et al., 2008; BATAGHIN et al., 2010), these numbers are related to the phytophysiognomic characteristics of the study area. The presence of abiotic factors that are more relevant to epiphytism, such as greater water availability (ZOTZ; HIETZ, 2001), means that epiphytes have their centers of diversity in moist forests (GENTRY; DODSON, 1987a). In addition, the greater availability of water in the Dense Ombrophilous Forest means that epiphytes can achieve a more elaborate niche partitioning, anchoring themselves to different parts of the forophytes (NIEDER et al. 1999).

The concentration of species in the Bromeliaceae, Orchidaceae and Polypodiaceae families, which is common in environments with greater humidity (KERSTEN, 2006; PETEAN 2009; BLUM, 2010), is related to the environmental factors characteristic of the core site, such as the less accentuated climatic variations and the absence of periods of water deficit. In addition, the Orchidaceae family shows a lower number of species and abundance than expected, a fact that allows us to infer possible anthropogenic interference in the area, especially in the collection of species of ornamental/economic interest, such as orchids.

The distribution of epiphytes in the forophyte strata at Sitio Core (Figure 41) showed that the base of the canopy was the stratum with the highest epiphyte abundance, with an abundance value (VA) of 642. The second most abundant stratum was the high canopy with VA of 439, followed by the inner canopy with VA = 380, the middle canopy with VA = 305, the low canopy with VA = 115 and the outer canopy with VA = 27.

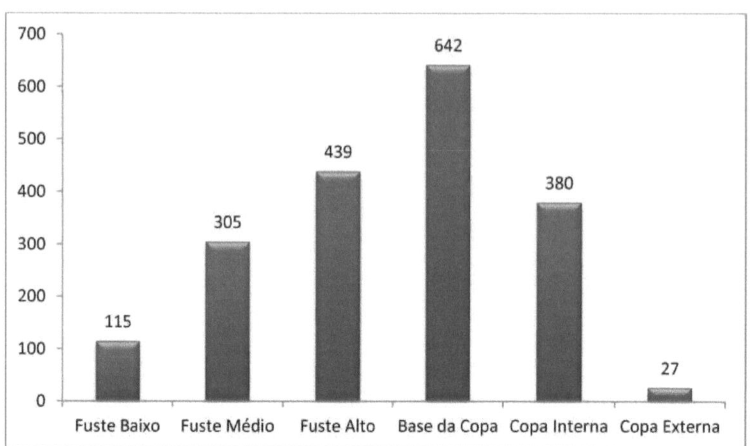

Figure 41: Distribution of the abundance of epiphytic vascular species among the forophytic strata in the Core Site of the Upstream Area in the Sorocaba/Médio Tietê Hydrographic Basin.

The statistical analysis applied to the vertical distribution of vascular epiphytes in the forophytic strata, based on species abundance, revealed a significant difference between the low stem and all the other strata (Table 41). This same pattern was observed for the outer canopy. The low stem also differed significantly from the base of the canopy (in addition to the low stem and outer canopy). This variation in abundance at the ends of the forophyte was to be expected given that the intermediate regions present better conditions for epiphyte development.

What is striking about this forest fragment is the reduced number of individuals in the outer canopy (Figure 41), which was significantly lower than that observed in the low stem, the first time this has happened in this study, including the results observed at the other sites in the watershed (sites in the Downstream and Central Areas). This is a reflection of the climatic conditions and also of the type of forest involved (FOD), however, in conserved forest areas such as this site (and also the Replica III site in the Downstream Area), the better microclimatic conditions in the low stem allow for the more abundant development of species that are demanding in terms of humidity, justifying the greater abundance in this region.

This can be considered an indication of the state of conservation of the forest fragments, because even in places with different climates (Downstream Area and Upstream Area), the epiphytes showed similar behavior in terms of vertical distribution on the forophytes, varying according to the degree of alteration of the remnant. In addition, in conserved fragments, among the species that occur in the low stem, those with greater plasticity are also recorded,

130

as are those that occur in the outer canopy, but there seems to be greater competition between species, which tends to reduce the number of individuals of generalist species, and increase the abundance of species characteristic of the more shaded regions. The lower abundance of the outer canopy, in turn, may mean that the species in this region are limited by factors such as temperature variation, which in this location is lower during the winter.

Table 39: Analysis of the distribution of the richness and abundance of vascular epiphytes among the forophytic strata in the Core Site of the Montante Area in the Sorocaba/Médio Tietê basin.

^^^-^^Richness Abundance^-^^	Beam Bass	Beam Medium	Beam High	World base	Cup Internal	Cup External
Low stem		0,05306	**1,71E-05**	**3,60E-07**	**0,0027**	**0,003612**
Medium stem	**0,004**		**0,01428**	**9,72E-04**	0,2765	**2,30E-06**
High stem	**7,77E-04**	0,126		0,3688	0,1677	**6,33E-12**
World Cup base	**1,18E-04**	**0,014**	0,114		**0,0239**	**3,74E-14**
Internal Cup	**0,002**	0,249	0,327	0,055		**1,18E-08**
External Cup	**8,31E-04**	**0,0001**	**0,0001**	**0,0001**	**0,0001**	

Species richness was significantly different (and lower) in the outer canopy than in all the other strata. This is indicative of the select group of species capable of surviving the more adverse microclimatic conditions of the upper part of the forest. Species composition was also significantly higher in the lower forest strata (low stem and medium stem) than in all strata (except for the medium stem and inner canopy - Table 39). This is because, in this site, it is possible to observe the occurrence of three groups of species: a) a more generalist group where almost all the species occur from the canopy to the ground; b) the group that is more demanding of moisture acquisition conditions, which is restricted to the lower strata, usually low and medium stem, sometimes reaching the base of the canopy; and c) a third group that occurs in the intermediate regions of the forophytes, especially at the base of the canopy and inner canopy, characterized by those species that need better conditions for anchoring propagules, or even demanding in terms of the presence of suspended soil which increases humidity even though they have better access to light. This corroborates what was proposed by Benzing (1990), who points out that the vertical evolution of vascular epiphytes occurred (and still occurs) through the exchange of more restrictive spaces for the acquisition of water in exchange for better light conditions.

Analysis of the vascular epiphytes at Sitio Règlica I in the upstream area of the Sorocaba/Médio Tietê basin

The survey of vascular epiphytes at Sitio Règlica I in the upstream area of the hydrographic basin was carried out on a private rural property in the municipality of Mairinque - SP. The forest remnant of approximately 260 ha, located at UTM coordinates 275.857 and 7.386.551 in zone 23 South, is characterized by being an ecotone area between the Semideciduous Seasonal Forest and the Dense Ombrophilous Forest. The steep slope of the area in the forest remnant played a fundamental role in maintaining the vegetation cover. Indications of anthropic presence were observed, such as the existence of trails (Figure 42) for the removal of wood, possibly also used for hunting, since "cevas" were observed, which are areas where hunters deposit food for the animals and then return to the hunting activity itself. The surrounding areas are characterized by anthropic use, with the presence of houses and vegetable plantations.

Figura 42: Registration of the existing trail at Sitio Règlica I in the Area Montante of the Sorocaba/Mèdio Tietê hydrographic basin.

The floristic analysis of the area revealed the presence of 37 species, belonging to 23 genera and eight families (Table 40). The Shannon index for the area was H' = 2.728, the equability (J) was 0.756 and the Margalef richness (d) was 5.141.

Table 40 - List of vascular epiphyte species found at Sitio Règlica I in the Montante area (Fazenda Dona Lùcia) of the Sorocaba/Médio Tietê hydrographic basin and their respective Ecological Categories (EC) - HLC: Holoepiphyte characteristic; HLF: Holoepiphyte facultative; HLA: Holoepiphyte accidental; HMP: Hemiepiphyte primary; HMS: Hemiepiphyte secondary. Dispersal form (Disp.) - Zo:= Zoochoric; An: Anemochoric. Reg: HUFSCar herbarium registration number (Im: Digital image).

N	Family	Species	EC	Disp.	Reg.
1	Aspleniaceae	Asplenium pteropus Kaulf.	HLC	An	8481
2	Blechnaceae	Blechnum binervatum (Poir.) C.V. Morton & Lellinger	HMS	An	8501
3	Bromeliaceae	Aechmea bromeliifolia (Rudge) Baker	HLC	Zo	8432
4	Bromeliaceae	Aechmea distichantha Lem.	HLC	Zo	8503
5	Bromeliaceae	Aechmea nudicaulis (L.) Griseb.	HLC	Zo	8434
6	Bromeliaceae	Billbergia distachya (Vell.) Mez	HLC	Zo	8445
7	Bromeliaceae	Billbergia zebrina (Herb.) Lindl.	HLC	Zo	8433
8	Bromeliaceae	Tillandsia araujei Mez	HLC	An	8377
9	Bromeliaceae	Tillandsia sp.	HLC	An	8425
10	Bromeliaceae	Tillandsia tenuifolia L.	HLC	An	8509
11	Bromeliaceae	Vriesea carinata Wawra	HLC	An	8513
12	Dryopteridaceae	Elaphoglossum ornatum (Mett. ex Kuhn) Christ	HLF	An	8476
13	Dryopteridaceae	Polybotrya cylindrica Kaulf.	HMS	An	8477
14	Orchidaceae	Campylocentrum aromaticum Barb.Rodr.	HLC	An	8419
15	Orchidaceae	Campylocentrum grisebachii Cogn.	HLC	An	Im
16	Orchidaceae	Capanemia micromera Barb. Rodr.	HLC	An	8423
17	Orchidaceae	Gomesa recurva R. Br.	HLC	An	8410
18	Orchidaceae	Gomesa glaziovii Cogn.	HLC	An	8411
19	Orchidaceae	Dichaea trulla Rchb. f.	HLC	An	8470
20	Orchidaceae	Lophiaris pumila (Lindl.) Braem	HLC	An	8468
21	Orchidaceae	Miltonia sp.	HLC	An	Im
22	Orchidaceae	Saundersia mirabilis Rchb.f.	HLC	An	8494
23	Piperaceae	Peperomia catharinae Miq.	HLC	Zo	8422
24	Piperaceae	Peperomia urocarpa Fisch. & C.A. Mey.	HLC	Zo	8492
25	Polypodiaceae	Campyloneurum acrocarpon Fée	HLC	An	8429
26	Polypodiaceae	Campyloneurum nitidum (Kaulf.) C. Presl	HLC	An	8531
27	Polypodiaceae	Ceradenia albidula (Baker) L.E. Bishop	HLC	An	8376
28	Polypodiaceae	Microgramma persicariifolia (Schrad.) C. Presl	HLC	An	8520
29	Polypodiaceae	Microgramma squamulosa (Kaulf.) de la Sota	HLC	An	8524
30	Polypodiaceae	Pleopeltis hirsutissima (Raddi) de la Sota	HLC	An	8521
31	Polypodiaceae	Pleopeltis macrocarpa (Bory ex Willd.) Kaulf.	HLC	An	8435

Keep going...

Table 40 - Continuation...

N	Family	Species	EC	Disp.	Reg.
32	Polypodiaceae	*Pleopeltis pleopeltifolia* (Raddi) Alston	HLC	An	8525
33	Polypodiaceae	*Serpocaulon catharinae* (Langsd. & Fisch.) A.R. Sm.	HLC	An	8528
34	Polypodiaceae	*Serpocaulon latipes* (Langsd. & Fisch.) A.R. Sm.	HLC	An	8423
35	Polypodiaceae	*Serpocaulon sehnemii* (Pic.-Serm.) Labiak & J.Prado	HLC	An	8532
36	Pteridaceae	*Polytaenium cajenense* (Desv.) Benedict	HLC	An	8439
37	Pteridaceae	*Vittaria lineata* (L.) Sm.	HLC	An	8454

Although there are no surveys in ecotone regions similar to the study area, the richness of vascular epiphytes at this site can be considered low when compared to studies carried out in dense ombrophilous forest by Blum (2010) - 277 species; Petean (2009) - 159; Kersten (2006) - 349 species; Breier (2005) - 161 species; Petean (2003) - 97 species; Schütz-Gatti (2000) - 175 species; Fontoura et al. (1997) - 293 species; and Hertel (1950) - 101 species. It is also lower than the richness observed in Semideciduous Seasonal Forest by Rogalski and Zanin (2003) - 70 species; Giongo and Waechter (2004) - 57 species; and Cervi and Borgo (2007) - 56 species. However, it is similar to the results obtained by Dislich and Mantovani (1998) - 34 species and Bonnet et al. (2011) - 35 species (average of six collection seasons), and is higher than the data from Borgo et al. (2002) - 32 species; Breier (2005) - 25 species; Dettke et al. (2008) - 29 species; and Bataghin et al. (2010) - 21 species.

Sitio Rèplica I showed greater richness than that found in Sitios Réplicas II and III, and less richness than that observed in Sitio Core in the upstream area of the Sorocaba/Médio Tietê basin. The number of epiphytic species recorded may be related to the complexity of the forest, either due to the favorable microclimate or even the number of available forophytes, especially larger trees, since these are factors that can affect epiphytic diversity (ENGWALD et al., 2000; BARTHLOTT et al., 2001; DETTKE et al., 2008).

The Polypodiaceae family was the richest in replicate site I, with 11 species. The Orchidaceae and Bromeliaceae families had nine species each, accounting for almost 50% of the vascular epiphytic diversity in this area. The Dryopteridaceae, Piperaceae and Pteridaceae families had two species each.

Characteristic holoepiphytes predominated at this site, accounting for over 91% of the species, followed by secondary hemiepiphytes with 5.4% of the species, and facultative holoepiphytes with 2.7% of the species. No accidental holoepiphytes or primary hemiepiphytes were found in the area. Characteristic holoepiphytes have been dominant in dense ombrophilous forests (BLUM, 2010; PETEAN, 2009; KERSTEN, 2006; BREIER, 2005; FONTOURA et al, 1997; SCHÜTZ-GATTI, 2000; PETEAN, 2003) and also in Semideciduous Seasonal Forest (PINTO et al., 1995; DISLICH; MANTOVANI, 1998; ROGALSKI; ZANIN, 2003; CERVI; BORGO, 2007; DETTKE et al., 2008; BATAGHIN et al., 2010).

At this site, the predominant dispersal syndrome was anemochoric with 81% of the species (30 spp.) and zoochoric with only seven species (19%). The higher number of anemochoric species compared to zoochoric species has been observed in several studies in Brazil (BREIER, 2005; DETTKE et al., 2008; MENINI-NETO et al., 2009; GERALDINO et al., 2010), and appears to be a trend for the Upstream Area of the watershed, especially in areas where the forest is more conserved and maintains greater structural complexity.

All 37 species were recorded in the quantitative analysis, with a Polypodiaceae being the most prominent species at Sitio Réplica I (Table 41).

Table 41 - Vascular epiphytes from Sitio Réplica I in the Montante area (Fazenda Dona Lùcia) of the Sorocaba/Médio Tietê hydrographic basin, classified according to the value of epiphytic importance - nr: absolute number of occurrences in the strata; far: absolute frequency in the strata; ni: absolute number of occurrences in the forophytic individuals; fai: absolute frequency in the forophytic individuals; vt (total value): sum of the abundance estimates; vie: epiphytic importance value; nota: average score obtained.

Species	nr	far	ni	fai	vt	vie	note
Microgramma squamulosa	171	31.7	54	60.0	286	25.98	1.67
Gomesa recurva	82	15.2	35	38.9	145	13.17	1.77
Pleopeltis hirsutissima	93	17.2	37	41.1	142	12.90	1.53
Pleopeltis pleopeltifolia	60	11.1	27	30.0	91	8.27	1.52
Peperomia urocarpa	28	5.2	19	21.1	49	4.45	1.75
Campyloneurum acrocarpon	21	3.9	11	12.2	33	3.00	1.57
Serpocaulon latipes	16	3.0	9	10.0	31	2.82	1.94
Lophiaris pumila	15	2.8	9	10.0	25	2.27	1.67
Campyloneurum nitidum	12	2.2	8	8.9	22	2.00	1.83
Saundersia mirabilis	15	2.8	8	8.9	21	1.91	1.40
Billbergia zebrina	11	2.0	8	8.9	18	1.63	1.64
Serpocaulon catharinae	10	1.9	4	4.4	18	1.63	1.80
Dichaea trulla	11	2.0	8	8.9	17	1.54	1.55
Miltonia sp.	11	2.0	7	7.8	17	1.54	1.55
Campylocentrum cf. grisebachii	13	2.4	7	7.8	16	1.45	1.23

Keep going...

Table 41 - Continuation

Species	nr	far	ni	fai	vt	vie	note
Elaphoglossum ornatum	12	2.2	7	7.8	16	1.45	1.33
Serpocaulon sehnemii	11	2.0	7	7.8	16	1.45	1.45
Billbergia distachya	7	1.3	4	4.4	15	1.36	2.14
Campylocentrum aromaticum	10	1.9	5	5.6	14	1.27	1.40
Pleopeltis macrocarpa	9	1.7	4	4.4	14	1.27	1.56
Vittaria lineata	6	1.1	2	2.2	12	1.09	2.00
Polytaenium cajenense	9	1.7	5	5.6	11	1.00	1.22
Microgramma persicariifolia	4	0.7	2	2.2	10	0.91	2.50
Aechmea bromeliifolia	4	0.7	2	2.2	9	0.82	2.25
Peperomia catharinae	4	0.7	2	2.2	9	0.82	2.25
Gomesa glaziovii	6	1.1	3	3.3	8	0.73	1.33

Species						
Tillandsia araujei	4	0.7	2	2.2	6	0.54	1.50
Asplenium pteropus	5	0.9	4	4.4	5	0.45	1.00
Aechmea distichantha	2	0.4	1	1.1	5	0.45	2.50
Polybotrya cylindrica	3	0.6	2	2.2	3	0.27	1.00
Blechnum binervatum	2	0.4	1	1.1	3	0.27	1.50
Tillandsia sp.	2	0.4	1	1.1	3	0.27	1.50
Tillandsia tenuifolia	2	0.4	1	1.1	3	0.27	1.50
Capanemia micromera	2	0.4	1	1.1	2	0.18	1.00
Ceradenia albidula	2	0.4	1	1.1	2	0.18	1.00
Aechmea nudicaulis	1	0.2	1	1.1	2	0.18	2.00
Vriesea carinata	1	0.2	1	1.1	2	0.18	2.00

Microgramma squamulosa (Polypodiaceae), had an epiphytic importance value (EVI) of 25.98 and an average score of 1.67, occurring in 60% of the forophytes and 31.7% of the strata, making it the most important species in this site. *Gomesa recurva* (Orchidaceae) had a VIE of 13.17 and an average score of 1.77, and was recorded in almost 40% of the forophytes and 15.2% of the strata. *Pleopeltis hirsutissima* (Polypodiaceae) was the third most important species on the site, recorded in 41.1% of the forophytes and 17.2% of the strata, with an average score of 1.53 and a VIE of 12.90 (Figure 43). Another Polypodiaceae - *Pleopeltis pleopeltifolia* - with a VIE of 8.27 and an average score of 1.52 was the fourth most important species at Sitio Réplica I. These four species were responsible for more than 60% of the VIE of this site.

Figura 43: Pleopeltis hirsutissima (Raddi) de la Sota (Polypodiaceae), the third most important epiphytic species at Sitio Rèplica I in the Àrea Montante.

The presence of a greater number of species, some with high abundance, meant that the Polypodiaceae family, responsible for more than 60% of the epiphytic importance value, was considered the most important in the site. The Orchidaceae family, which had nine species, was the second most important on the site with a VIE of 24.07. Bromeliaceae and Piperaceae, which had nine and two species respectively, had a VIE of 5.72 and 5.27, and were the third and fourth most important families in Sitio Règplica I da Àrea Montante. Although the concentration of abundance in a few species is typical of impacted areas (BATAGHIN et al., 2010) and may be related to models of pre-emptying of niches (MAY, 1975), it is important to note that, at this site, this fact may be related to the characterization of the forest's phytophysiognomic type. In other words, the great richness of Polypodiaceae, Orchidaceae and Bromeliaceae, with a high abundance of the first family, may be an indication that the forest studied has characteristics more related to a Semideciduous Seasonal Forest and not to a Dense Ombrophilous Forest. Kersten (2006, 2009) highlights the high importance of the Polypodiaceae family in areas of Seasonal Forest, with less prominence in areas of Dense Ombrophilous Forest.

The distribution of epiphytes in the forophyte strata (Figure 44) showed that the base of the canopy, with an abundance value (VA) of 342, was the stratum with the highest epiphyte abundance. The second most abundant stratum was the high stem, with VA = 263, followed by the inner crown, with VA = 187, the middle stem, with VA = 168, the low stem, with VA = 108 and the outer crown, with VA = 33.

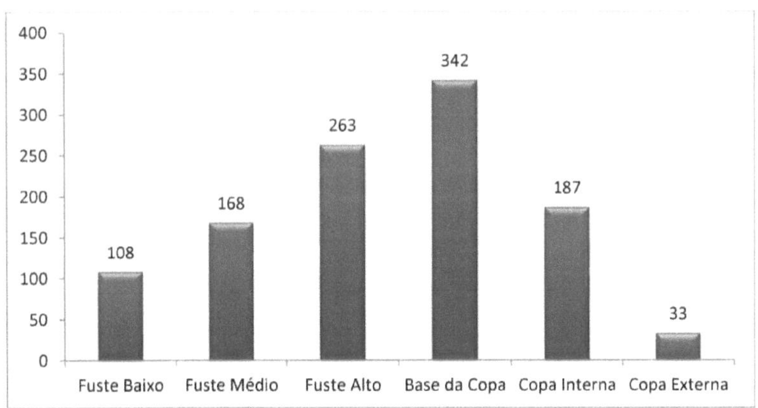

Figura 44: Distribution of the abundance of epiphytic vascular species among the forophytic strata at Sitio Réplica I in the Montante area of the Sorocaba/Médio Tietê Hydrographic Basin.

The distribution of vascular epiphytes in the forophytic strata, based on the abundance of species, showed that the outer canopy differed significantly from all the other strata. In addition to the outer canopy, the low canopy differed significantly from the high canopy and the base of the canopy (Table 42). The lowest number of individuals was observed in the outer canopy (Figure 44) and, following the same pattern as the core site of the Upland Area, this may indicate the limitation of this region to epiphytic individuals and species.

The greater abundance of the low stem in relation to the outer canopy can be attributed to the climate of the geographical region (predominantly humid). Evidence of the importance of the general climatic conditions of the region on the epiphytic community of this site is due to the fact that even though the forest fragment shows signs of anthropogenic disturbance, the lower strata are capable of harboring a considerable number of epiphytic specimens.

Table 42: Analysis of the distribution of the richness and abundance of vascular epiphytes among the forophytic strata at Sitio Réplica I in the Montante Area of the Sorocaba/Médio Tietê hydrographic basin.

^^^-^^^Richness Abundance^'^^^	Beam Bass	Beam Medium	Beam High	World base	Cup Internal	Cup External
Low stem		**0,0095**	**0,0095**	**0,0190**	0,995	**2,80E-04**
Medium stem	0,143		0,9942	0,7971	**0,0095**	**5,09E-09**
High stem	**0,041**	0,136		0,7971	**0,0095**	**5,09E-09**
World Cup base	**0,032**	0,079	0,287		**0,0190**	**1,72E-08**
Internal Cup	0,231	0,428	0,273	0,156		**2,80E-04**
External Cup	**0,019**	**0,002**	**0,0003**	**0,0006**	0,001	

Another important aspect observed for this site (and also for the Core Site in the Upland Area) is a shift in abundance down the vertical gradient. In almost all the sites in the Downstream and Central Areas, the inner canopy had a greater number of individuals than the high stem, but this pattern is reversed in these two sites in the Upland Area. This may be a consequence of the phytophysiognomic characteristics of the forest (wetter and shadier in this part of the basin), but it may also indicate a greater sensitivity of the epiphytic community to environmental conditions in the upper strata (inner and outer canopy) of the forest.

In terms of richness, the greater restriction of species to the upper strata can also be seen, as the outer canopy has a significantly different community from all the strata, as well as a reduced number of species (only three). The differences in the low stem and inner canopy are also due to the reduction in the number of species, although there are distinct species in these two strata, they share a large number of species. In addition, the greater number of species and individuals in the lower canopy is a characteristic of better conserved areas.

Analysis of the vascular epiphytes at Sitio Réplica II in the upstream area of the Sorocaba/Médio Tietê basin

Sitio Réplica II of the Area Montante of the Sorocaba/Médio Tietê watershed was established in a forested area of a private property in the municipality of Sao Roque - SP. The forest fragment, located at UTM coordinates 284.361 and 7.389.111 in zone 23 South, is a forest remnant characterized as an ecotone between Semideciduous Seasonal Forest and Dense Ombrophilous Forest. The area has approximately 321 ha and an altitude of around 1050 meters. The area's surroundings are characterized by anthropogenic use, with the presence of houses, cattle and horse breeding, as well as fruit and vegetable plantations, especially tomato and artichoke plantations, which rely on the constant application of pesticides, which can influence the fragment's vascular epiphytic community. In this area, the undergrowth is reduced, possibly due to the presence of several trails, including those with a continuous flow of humans, as well as traps used for hunting wild species (Figure 45).

Figura 45: Snare traps (illegal capture of wild animals) present at the Réplica II site in the upstream area.

At Sitio Réplica II in the Area Montante in the Sorocaba/Médio Tietê basin, 35 species were found, belonging to 23 genera and 10 families (Table 43). The site's Shannon index was H' = 2.713, equability (J) was 0.769 and Margalef's richness (d) was 4.765.

Table 43 - List of vascular epiphyte species found at Sitio Réplica II in the Montante area (Fazenda Dona Antonia) of the Sorocaba/Médio Tietê hydrographic basin and their respective Ecological Categories (EC). HLC: Holoepiphyte characteristic; HLF: Holoepiphyte facultative; HLA: Holoepiphyte accidental; HMP: Hemiepiphyte primary; HMS: Hemiepiphyte secondary. Dispersal form (Disp.) - Zo: Zoochoric; An: Anemochoric. Reg. = HUFSCar Deposit Number (Im = Digital Image).

N	Family	Species	EC	Disp.	Reg.
1	Araceae	Anthurium sellowianum Kunth	HLF	An	Im

2	Blechnaceae	Blechnum binervatum (Poir.) C.V. Morton & Lellinger	HMS	An	8501
3	Bromeliaceae	Aechmea distichantha Lem.	HLC	Zo	8503
4	Bromeliaceae	Billbergia distachya (Vell.) Mez	HLC	Zo	8445

Keep going...

Table 43 - Continuation...

N	Family	Species	EC	Disp.	Reg.
5	Bromeliaceae	*Billbergia zebrina* (Herb.) Lindl.	HLC	Zo	8433
6	Bromeliaceae	*Tillandsia araujei* Mez	HLC	An	8377
7	Bromeliaceae	*Tillandsia geminiflora* Brongn.	HLC	An	8427
8	Bromeliaceae	*Tillandsia recurvata* (L.) L.	HLC	An	8429
9	Cactaceae	*Rhipsalis teres* (Vell.) Steud.	HLC	Zo	8384
10	Cactaceae	*Rhipsalis trigona* Pfeiff.	HLC	Zo	8381
11	Commelinaceae	*Tradescantia albiflora* Kunth	HLA	Zo	8391
12	Dryopteridaceae	*Elaphoglossum ornatum* (Mett. ex Kuhn) Christ	HLF	An	8476
13	Orchidaceae	*Campylocentrum aromaticum* Barb.Rodr.	HLC	An	8419
14	Orchidaceae	*Capanemia micromera* Barb. Rodr.	HLC	An	8423
15	Orchidaceae	*Coppensia varicosa* (Lindl.)Campacci	HLC	An	8406
16	Orchidaceae	*Cyclopogon multiflorus* Schltr.	HLA	An	8404
17	Orchidaceae	*Lophiaris pumila* (Lindl.) Braem	HLC	An	8468
18	Orchidaceae	*Rodriguezia decora* (Lem.) Rchb.f.	HLC	An	8441
19	Orchidaceae	*Saundersia mirabilis* Rchb.f.	HLC	An	8494
20	Piperaceae	*Peperomia alata* Ruiz & Pav.	HLC	An	8420
21	Piperaceae	*Peperomia castelosensis* Yunck.	HLC	Zo	8421
22	Piperaceae	*Peperomia glabella* (Sw.) A. Dietr.	HLF	Zo	8475
23	Piperaceae	*Peperomia trineura* Miq.	HLC	Zo	8446
24	Polypodiaceae	*Campyloneurum acrocarpon* Fée	HLC	An	8529
25	Polypodiaceae	*Campyloneurum nitidum* (Kaulf.) C. Presl	HLC	An	8531
26	Polypodiaceae	*Microgramma persicariifolia* (Schrad.) C. Presl	HLC	An	8520
27	Polypodiaceae	*Microgramma squamulosa* (Kaulf.) de la Sota	HLC	An	8534
28	Polypodiaceae	*Pecluma* sp.	HLC	An	8375
29	Polypodiaceae	*Pleopeltis hirsutissima* (Raddi) de la Sota	HLC	An	8521
30	Polypodiaceae	*Pleopeltis pleopeltifolia* (Raddi) Alston	HLC	An	8525
31	Polypodiaceae	*Pleopeltis squalida* (Vell.) de la Sota	HLC	An	8436
32	Polypodiaceae	*Serpocaulon fraxinifolium* (Jacq.) A.R. Sm.	HLC	An	8518
33	Polypodiaceae	*Serpocaulon latipes* (Langsd. & Fisch.) A.R. Sm	HLC	An	8423
34	Pteridaceae	*Polytaenium cajenense* (Desv.) Benedict	HLC	An	8439
35	Pteridaceae	*Vittaria lineata* (L.) Sm.	HLC	An	8454

The richness of vascular epiphytes at this site can be considered low when compared to studies carried out in Dense Ombrophilous Forest by Blum (2010) - 277 species; Petean (2009) - 159 species; Kersten (2006) - 349 species; Breier (2005) - 161 species; Petean (2003) - 97 species; Schütz-Gatti (2000) - 175 species; Fontoura *et al.* (1997) - 293 species; and

Hertel (1950) - 101 species, and also lower than that observed in Semideciduous Seasonal Forest by Rogalski and Zanin (2003) - 70 species; Giongo and Waechter (2004) - 57 species and Cervi and Borgo (2007) - 56 species. However, it is similar to the results obtained by Dislich and Mantovani (1998) - 34 species, Bonnet et al. (2011) - 35 species (average of six collection seasons) and Borgo et al. (2002) - 32 species; but it is higher than the data from Breier (2005) - 25 species; Dettke et al. (2008) - 29 species; and Bataghin et al. (2010) - 21 species.

Replica site II showed a lower richness than that observed in the Core Site of the Upstream Area (PEJU) and similar to that found in Replica Site I sampled in this study area. The presence and abundance of species from the Polypodiaceae family may be indicative of the phytophysiognomic condition of the forest fragment, especially due to the occurrence of species from the Microgramma and Pleopeltis genera, which are characterized as pioneers (BONNET et al., 2011), 2011) and are recorded with greater frequency and abundance in Semideciduous Seasonal Forest formations (KERSTEN; KUNIYOSHI, 2009; BONNET et al., 2009; BATAGHIN et al., 2010). Dettke et al. (2008) and Bataghin et al. (2010) also point out that species of these genera are characteristic of areas with some degree of anthropogenic interference, and this is a possibility for the area of Sitio Réplica II da Area Montante, given the presence of trails, with people passing through, as well as garbage being deposited on the site.

The Polypodiaceae family, cited by Kersten (2006) as one of the most important for seasonal forest areas, was the richest at this site, with 10 species. Bromeliaceae and Orchidaceae had six species each and ranked second in richness. Piperaceae had four species and Cactaceae and Pteridaceae had two species each. The three main families were responsible for almost 65% of the site's epiphytes. Characteristic holoepiphytes were predominant, accounting for over 91% of the species, followed by facultative holoepiphytes which accounted for 9% of the species; no accidental holoepiphytes or primary or secondary hemiepiphytes were found. Characteristic holoepiphytes have been dominant in dense ombrophilous forests (BLUM, 2010; PETEAN, 2009; KERSTEN, 2006; BREIER, 2005; FONTOURA et al, 1997; SCHÜTZ-GATTI, 2000; PETEAN, 2003) and also in Semideciduous Seasonal Forest (PINTO et al., 1995; DISLICH; MANTOVANI, 1998; ROGALSKI; ZANIN, 2003; CERVI; BORGO, 2007; DETTKE et al., 2008; BATAGHIN et al., 2010).

At this site, 75% of the species had an anemochoric dispersal syndrome (26 spp.) and around 25% had a zoochoric dispersal syndrome (nine species). The recording of a greater number of anemochoric species than zoochoric species has been observed in several studies in Brazil

(BREIER, 2005; DETTKE et al., 2008; MENINI-NETO et al., 2009; GERALDINO et al., 2010), and is proving to be a trend in the upstream area of the watershed, especially in areas where the forest is more conserved and maintains greater structural complexity and, consequently, a microclimate favorable to epiphytism.

In the quantitative analysis, 34 species were recorded, with three Polypodiaceae being the most prominent species at Sitio Réplica II (Table 44). *Microgramma squamulosa had* an epiphytic importance value (VIE) of 27.80 and an average score of 1.72, occurring in almost 56% of the forophytes and 30.6% of the strata, making it the most important species at this site. *Pleopeltis pleopeltifolia* had a VIE of 10.22, an average score of 1.44, and was recorded in 27.8% of the forophytes and 13.3% of the strata. *Pleopeltis hirsutissima* had a VIE of 7.66 and an average score of 1.66, and was recorded in 20.0% of the forophytes and 8.7% of the strata.

Table 44 - Vascular epiphytes from Sitio Réplica II in the Montante area (Fazenda Dona Antonia) of the Sorocaba/Médio Tietê hydrographic basin, classified according to the value of epiphytic importance - nr: absolute number of occurrences in the strata; far: absolute frequency in the strata; ni: absolute number of occurrences in the forophytic individuals; fai: absolute frequency in the forophytic individuals; vt (total value): sum of the abundance estimates; vie: epiphytic importance value; nota: average score obtained.

Species	nr	far	ni	fai	vt	vie	note
Microgramma squamulosa	165	30.6	50	55.6	283	27.80	1.72
Pleopeltis pleopeltifolia	72	13.3	25	27.8	104	10.22	1.44
Pleopeltis hirsutissima	47	8.7	18	20.0	78	7.66	1.66
Tillandsia geminiflora	54	10.0	22	24.4	77	7.56	1.43
Billbergia distachya	42	7.8	23	25.6	66	6.48	1.57
Anthurium sellowianum	27	5.0	14	15.6	46	4.52	1.70
Campyloneurum nitidum	27	5.0	14	15.6	46	4.52	1.70
Varicose coppensia	22	4.1	11	12.2	39	3.83	1.77
Campyloneurum acrocarpon	30	5.6	13	14.4	37	3.63	1.23
Serpocaulon latipes	25	4.6	11	12.2	35	3.44	1.40
Polytaenium cajenense	19	3.5	7	7.8	27	2.65	1.42
Pecluma sp.	15	2.8	7	7.8	25	2.46	1.67
Aechmea distichantha	13	2.4	8	8.9	21	2.06	1.62
Tillandsia recurvata	10	1.9	4	4.4	18	1.77	1.80
Rhipsalis teres	8	1.5	3	3.3	12	1.18	1.50
Pleopeltis squalida	6	1.1	2	2.2	11	1.08	1.83
Elaphoglossum ornatum	7	1.3	4	4.4	10	0.98	1.43

| *Peperomia castelosensis* | 6 | 1.1 | 3 | 3.3 | 10 | 0.98 | 1.67 |
| *Lophiaris pumila* | 5 | 0.9 | 3 | 3.3 | 7 | 0.69 | 1.40 |

Keep going...

Table 44 - Continuation...

Species	nr	far	ni	fai	vt	vie	note
Peperomia alata	4	0.7	2	2.2	7	0.69	1.75
Billbergia zebrina	4	0.7	2	2.2	6	0.59	1.50
Blechnum binervatum	4	0.7	2	2.2	6	0.59	1.50
Cyclopogon multiflorus	4	0.7	2	2.2	5	0.49	1.25
Microgramma persicariifolia	4	0.7	2	2.2	5	0.49	1.25
Tillandsia araujei	4	0.7	2	2.2	5	0.49	1.25
Peperomia trineura	3	0.6	1	1.1	5	0.49	1.67
Rhipsalis trigona	3	0.6	1	1.1	5	0.49	1.67
Peperomia glabella	2	0.4	1	1.1	5	0.49	2.50
Serpocaulon fraxinifolium	3	0.6	1	1.1	4	0.39	1.33
Campylocentrum aromaticum	2	0.4	1	1.1	3	0.29	1.50
Rodriguezia decorates	2	0.4	1	1.1	3	0.29	1.50
Saundersia mirabilis	2	0.4	1	1.1	3	0.29	1.50
Vittaria lineata	2	0.4	1	1.1	3	0.29	1.50
Capanemia micromera	1	0.2	1	1.1	1	0.10	1.00

Two species of Bromeliaceae are also worth mentioning: *Tillandsia geminiflora* (Figure 46) with a VIE of 7.56 and an average score of 1.43 and *Billbergia distachya with a* VIE = 6.48 and an average score of 1.57. These species contribute to the Bromeliaceae family being the second most important in the site, reaching 18.96% of the epiphytic importance value. The Polypodiaceae family, the most important, was responsible for 61.69% of the VIE. The Orchidaceae family, despite being one of the richest at the site, had the lowest relative abundance with a VIE of 5.5.

Figura 46: *Tillandsia geminiflora* Brongn. (Bromeliaceae), one of the most abundant

143

species at Sitio Réplica II in the Area Montante.

The concentration of richness and abundance in the Polypodiaceae and Bromeliaceae families is characteristic of areas with climatic conditions that are adverse to epiphytism, either due to the natural characteristics of the vegetation (BREIER, 2005) or to interference from anthropogenic activities (DETTKE et al., 2008; BATAGHIN et al., 2010).

Analysis of the forophytic strata revealed the base of the canopy (Figure 47) as the most abundant stratum, with an abundance value (AV) of 320, the high stem was second, with an AV of 262, followed by the inner canopy with an AV of 206. The middle stem, low stem and outer canopy had abundance values of 114, 59 and 57, respectively.

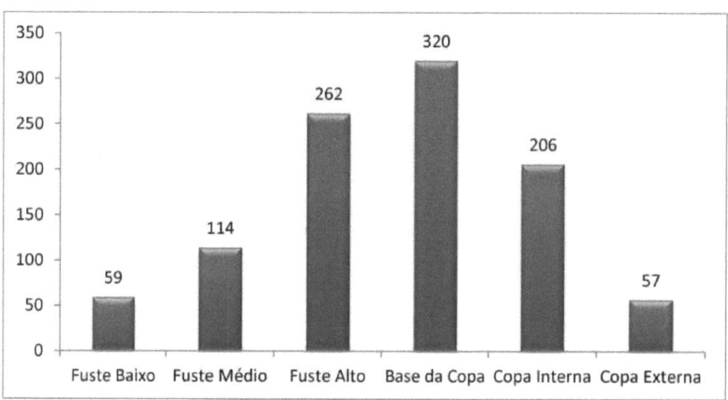

Figura 47: Distribution of the abundance of epiphytic vascular species among the forophytic strata at Sitio Réplica II in the Upstream Area of the Sorocaba/Médio Tietê Hydrographic Basin.

The analysis of the distribution of vascular epiphyte abundances in the forophytic strata showed the outer canopy to be different from all strata except the low stem. The high stem and the base of the canopy were significantly more abundant than the low and medium stems (Table 45). This model of abundance distribution is similar to that observed in the sites under anthropogenic influence in the Downstream and Central Areas, which applies to this forest fragment, since it showed various signs of human activity, including a trail along which there is frequent traffic, and a hunting trap was found in the densest part of the forest. Further evidence of this is the reduction in the number of individuals in the lower canopy, which in this site showed almost identical abundance to the outer canopy, a result similar to that observed in the most impacted forest fragments and different to that recorded in the Core and Replica I sites in the Montante area. This reinforces the idea that impacted forest areas show a

144

reduction in the number of individuals in the low stem, since atropic activities alter the microclimatic conditions of forest fragments (BATAGHIN et al., 2008; BARTHLOTT et al., 2001), especially by altering or removing the understory, which reduces the structural complexity of the forest.

Table 45: Analysis of the distribution of the richness and abundance of vascular epiphytes among the forophytic strata at Site Replica II of the Upland Area in the Sorocaba/Médio Tietê basin.

^^^-^^^Rich Abundance'~~~~~^	Beam Bass	Beam Medium	Beam High	World Cup base	Cup Internal	Cup External
Low stem		0,05458	**0,007582**	**0,007582**	0,1506	**0,01611**
Medium stem	0,0669		0,439	0,439	0,6255	**2,98E-05**
High stem	**0,0035**	**0,032**		0,9936	0,206	**1,29E-06**
World Cup base	**0,0011**	**0,025**	0,319		0,206	**1,29E-06**
Internal Cup	0,054	0,163	0,314	0,197		**1,85E-04**
External Cup	0,481	**1,33E-04**	**5,85E-06**	**3,74E-06**	**4,79E-04**	

Similar to what was observed in the anthropized sites in the Central and Downstream Areas, the richness of the outer canopy was reduced (five species), especially when compared to the low canopy (14 species), which was reflected in a significant difference between the outer canopy and all the others. The high stem and the base of the canopy also differed significantly from the low stem. These two strata (high stem and base of the canopy) had the highest number of species, indicating that they are a favorable environment for the development of epiphytes. However, the distribution of species on the forophytes observed at this site reflects the importance of anthropogenic factors on the vascular epiphytic community, since both the Core Site and the Replica I Site in the Upland Area have climatic conditions similar to those present in this forest remnant, suggesting that the differences in the vertical distribution of species are the result of environmental changes.

Analysis of the vascular epiphytes at Sitio Réplica III in the upstream area of the Sorocaba/Médio Tietê basin

The survey of vascular epiphytes at Sitio Réplica III in the upstream area of the Sorocaba/Médio Tiete river basin was carried out in an area of permanent preservation on the banks of the Itupararanga Reservoir, between the municipalities of Aluminio and Votorantim - SP. The forest remnant covers around 270 ha and is located between UTM coordinates 262.383 and 7.387.015 in zone 23 South, and is characterized by being a stretch of Semideciduous Seasonal Forest. At Sitio Réplica III, signs of anthropogenic influence were found, such as the presence of trails (Figure 48), an area used for camping with traces of campfires and garbage abandoned in the forest. This is possibly because the surrounding areas are characterized by the presence of a municipal road and the Itupararanga Reservoir, which attracts many fishermen to this location.

Figura 48:   Trail and camping site at Sitio Réplica III in the Area Montante.

Sitio Réplica III was the area with the lowest diversity of epiphytic species in the Upstream Area of the Sorocaba/Médio Tietê basin, with 28 species recorded, belonging to 19 genera and seven families (Table 46). The Shannon index for the area was H' = 2.364, the equability (J) was 0.709 and the Margalef richness (d) was 4.129.

Table 46 - List of vascular epiphyte species found at Sitio Réplica III in the Upstream Area (Itupararanga Reservoir) of the Sorocaba/Médio Tietê river basin and their respective Ecological Categories (EC) - HLC: Characteristic Holoepiphyte; HLF: Facultative Holoepiphyte; HLA: Accidental Holoepiphyte; HMP: Primary Hemiepiphyte; HMS: Secondary Hemiepiphyte. Dispersal form (Disp.) - Zo: Zoochoric; An: Anemochoric. Reg: HUFSCar herbarium registration number (Im: Digital image).

N	Family	Species	EC	Disp.	Reg.
1	Araceae	Anthurium sellowianum Kunth	HLF	Zo	Im
2	Aspleniaceae	Asplenium scandicinum Kaulf.	HLC	An	8482

Keep going...

Table 46- Continuation...

N	Family	Species	EC	Disp.	Reg.
3	Bromeliaceae	*Aechmea bromeliifolia* (Rudge) Baker	HLC	Zo	8432
4	Bromeliaceae	*Billbergia distachya* (Vell.) Mez	HLC	Zo	8445
5	Bromeliaceae	*Tillandsia recurvata* (L.) L.	HLC	An	8429
6	Bromeliaceae	*Tillandsia* sp.	HLC	An	8425
7	Bromeliaceae	*Vriesea gigantea* Mart. ex Schult. f.	HLC	An	8512
8	Dryopteridaceae	*Elaphoglossum ornatum* (Mett. ex Kuhn) Christ	HLF	An	8476

9	Orchidaceae	*Campylocentrum aromaticum* Barb.Rodr.	HLC	An	8419
10	Orchidaceae	*Gomesa* cf. *glaziovii* Cogn.	HLC	An	8411
11	Orchidaceae	*Gomesa* sp.	HLC	An	8417
12	Orchidaceae	*Lophiaris pumila* (Lindl.) Braem	HLC	An	8468
13	Orchidaceae	*Notylia longispicata* Hoehne & Schltr.	HLC	An	8416
14	Orchidaceae	*Polystachya estrellensis* Rchb.f.	HLC	An	8467
15	Orchidaceae	*Stelis deregularis* Barb. Rodr.	HLC	An	8473
16	Polypodiaceae	*Campyloneurum acrocarpon* Fée	HLC	An	8529
17	Polypodiaceae	*Campyloneurum repens* (Aubl.) C. Presl	HLC	An	8433
18	Polypodiaceae	*Campyloneurum nitidum* (Kaulf.) C. Presl	HLC	An	8531
19	Polypodiaceae	*Microgramma lycopodioides* (L.) Copel.	HLC	An	8437
20	Polypodiaceae	*Microgramma percussa* (Cav.) de la Sota	HLC	An	8392
21	Polypodiaceae	*Microgramma squamulosa* (Kaulf.) de la Sota	HLC	An	8534
22	Polypodiaceae	*Microgramma vacciniifolia* (Langsd. & Fisch.) Copel.	HLC	An	8522
23	Polypodiaceae	*Pecluma* sp.	HLC	An	8375
24	Polypodiaceae	*Pleopeltis hirsutissima* (Raddi) de la Sota	HLC	An	8521
25	Polypodiaceae	*Pleopeltis pleopeltifolia* (Raddi) Alston	HLC	An	8525
26	Polypodiaceae	*Serpocaulon fraxinifolium* (Jacq.) A.R.Sm.	HLC	An	8518
27	Polypodiaceae	*Serpocaulon latipes* (Langsd. & Fisch.) A.R.Sm.	HLC	An	8423
28	Pteridaceae	*Vittaria lineata* (L.) Sm.	HLC	An	8454

The richness of Sitio Réplica III can be considered low when compared to studies carried out in Semideciduous Seasonal Forest, such as Rogalski and Zanin (2003) who found 70 species, Giongo and Waechter (2004) who sampled 57 species, Cervi and Borgo (2007) who found 56 species and Bonnet et al. (2011) who found 60 species, which is similar to that observed by Dislich and Mantovani (1998), with 34 species; by Borgo et al. (2002), with 32 species; by Breier (2005), with 25 species, by Dettke et al. (2008), with 29 species; by Bataghin et al. (2010), with 21 species; and by Aguiar et al. (1981), who sampled 17 species. It is important to mention that Bonnet et al. (2011), studying several fragments of Semideciduous Seasonal Forest in the basin of the Tibagi River - PR (a similar study to the one carried out in the present study), revealed an average of 35 species per forest area, a similar number to the one found here. One factor that may have negatively influenced the diversity of this site is anthropic interference in the environment, since during data collection, people were often found passing through the fragment, usually for fishing purposes.

Sitio Réplica III, although it has the lowest diversity of vascular epiphytes among the sites in the Area Montante, plays an important role in protecting water resources. Several studies have reported that changes in the landscape negatively influence the diversity and abundance of vascular epiphytes (ENGWALD et al., 2000; BONNET; QUEIROZ, 2000; BARTHLOTT et al., 2001; BATAGHIN et al., 2008; DETTKE et al.., 2008), however, the conservation of forest remnants, especially those not protected by UCs, should not be overlooked, as the forest fragmentation of the Sorocaba/Médio Tietê basin may have isolated or even restricted the epiphytic community in these forest fragments.

The richest families at Sitio Réplica III were Polypodiaceae with 12 species and Orchidaceae with seven species. The Bromeliaceae family had five species. Araceae, Aspleniaceae, Dryopteridaceae and Pteridaceae had one species each. Characteristic holoepiphytes were dominant at the site, accounting for 26 species (93%), followed by facultative holoepiphytes (7%); no accidental holoepiphytes, primary hemiepiphytes or secondary hemiepiphytes were recorded. The predominance of characteristic holoepiphytes has also been observed in various studies in dense ombrophilous forest (BLUM 2010, PETEAN 2009, KERSTEN 2006, BREIER 2005, FONTOURA et al. 1997, SCHÜTZ-GATTI 2000, PETEAN 2003), as well as in Seasonal Semideciduous Forest (PINTO et al. 1995, DISLICH and MANTOVANI 1998, ROGALSKI and ZANIN 2003, CERVI and BORGO 2007, DETTKE et al. 2008, BATAGHIN et al. 2010).

Anemochory has predominated as a dispersal syndrome among epiphytic species (BENZING, 1987; BREIER, 2005; DETTKE et al., 2008; MENINI-NETO et al., 2009; GERALDINO et al.; 2010), which is corroborated by the results from this site, where almost 90% of the species showed anemochoric dispersal (25 spp.), while only 10% of the epiphytic species (3 spp.) showed zoochory as a dispersal syndrome. This high percentage of anemochory is a reflection of the large number of orchids and ferns recorded at this site in the Upland Area.

All 28 species found in the floristic survey were recorded in the quantitative analysis of Sitio Réplica III (Table 47).

Table 47 - Vascular epiphytes from Sitio Réplica III in the upstream area (Itupararanga Reservoir) of the Sorocaba/Médio Tietê hydrographic basin, classified according to the value of epiphytic importance - nr: absolute number of occurrences in the strata; far: absolute frequency in the strata; ni: absolute number of occurrences in the forophytic individuals; fai: absolute frequency in the forophytic individuals; vt (total value): sum of the abundance estimates; vie: epiphytic importance value; nota: average score obtained.

Species	nr	far	ni	fai	vt	vie	note
Microgramma squamulosa	201	37.2	58	64.4	331	31.92	1.65
Pleopeltis pleopeltifolia	135	25.0	49	54.4	193	18.61	1.43
Pleopeltis hirsutissima	102	18.9	44	48.9	144	13.89	1.41
Campyloneurum acrocarpon	41	7.6	20	22.2	54	5.21	1.32
Microgramma percussa	28	5.2	11	12.2	41	3.95	1.46

Billbergia distachya	20	3.7	9	10.0	33	3.18	1.65
Microgramma vacciniifolia	19	3.5	7	7.8	32	3.09	1.68
Elaphoglossum ornatum	18	3.3	10	11.1	25	2.41	1.39
Notylia longispicata	21	3.9	12	13.3	24	2.31	1.14
Serpocaulon latipes	16	3.0	8	8.9	24	2.31	1.50
Campyloneurum nitidum	16	3.0	8	8.9	23	2.22	1.44
Aechmea bromeliifolia	11	2.0	5	5.6	21	2.03	1.91
Pecluma sp.	7	1.3	3	3.3	12	1.16	1.71
Microgramma lycopodioides	7	1.3	2	2.2	11	1.06	1.57
Campylocentrum aromaticum	9	1.7	6	6.7	10	0.96	1.11
Tillandsia sp.	6	1.1	3	3.3	8	0.77	1.33
Gomesa glaziovii	4	0.7	2	2.2	7	0.68	1.75
Anthurium sellowianum	4	0.7	2	2.2	6	0.58	1.50
Campyloneurum repens	4	0.7	2	2.2	6	0.58	1.50
Vittaria lineata	4	0.7	2	2.2	6	0.58	1.50
Asplenium scandicinum	3	0.6	1	1.1	5	0.48	1.67
Gomesa sp.	3	0.6	2	2.2	4	0.39	1.33
Tillandsia recurvata	3	0.6	1	1.1	4	0.39	1.33
Vriesea gigantea	3	0.6	2	2.2	4	0.39	1.33
Serpocaulon fraxinifolium	2	0.4	1	1.1	3	0.29	1.50
Stelis deregularis	2	0.4	1	1.1	3	0.29	1.50
Lophiaris pumila	2	0.4	2	2.2	2	0.19	1.00
Polystachya estrellensis	1	0.2	1	1.1	1	0.10	1.00

Microgramma squamulosa (Polypodiaceae) had an epiphytic importance value (EVI) of 31.92 and an average score of 1.65, being present in 64.4% of the forophytes and 37.2% of the strata, making it the most important species in this site. *Pleopeltis pleopeltifolia* (Polypodiaceae), was the second most important species with a VIE of 18.61 and an average score of 1.43, as well as being present in 54.4% of the forophytes and 25% of the strata (Figure 49). The third most important species at this site was another

Polypodiaceae, Pleopeltis hirsutissima, with a VIE of 13.89 and a score of 1.41, was recorded in 48.9% of the forophytes and 18.9% of the strata.

Figura 49: Pleopeltis pleopeltifolia (Raddi) Alston (Polypodiaceae), the second most abundant species at the Réplica III site in the Montante Area.

At the Replica III site in the upstream area, the Polypodiaceae family was the most important, with over 84% of the epiphytic importance value. The Bromeliaceae family was the second most important with a total VIE of 6.75, followed by Orchidaceae with a VIE of 4.92. Although the Polypodiaceae, Bromeliaceae and Orchidaceae are among the richest epiphyte families in the world (GENTRY; DODSON, 1987b), and in Brazil (KERSTEN, 2006), the greater diversity and, above all, the high abundance of species in the Polypodiaceae family - a family that tends to be more expressive in areas under anthropogenic pressure (DETTKE et al., 2008; BATAGHIN et al., 2010) or in areas with a microclimate that is not very favorable to epiphytism (BERNARDI; BUDKE, 2009) - indicate an environment that is not very favorable to the establishment and development of vascular epiphytes. It is important to note that there is a dominance of species in relation to the distribution of epiphytic importance values at this site (Table 47), which suggests the existence of an imbalance in the epiphytic community, indicative of the compromised state of conservation of the forest area at this site.

With regard to the vertical distribution of epiphytes (Figure 50), the base of the canopy was the forophytic stratum with the highest abundance (VA = 322), followed by the high stem, the second most abundant stratum with VA of 293, and the inner canopy with VA of 202. The middle canopy had an abundance value of 125, while the low canopy and outer canopy had abundance values of 48 and 47, respectively.

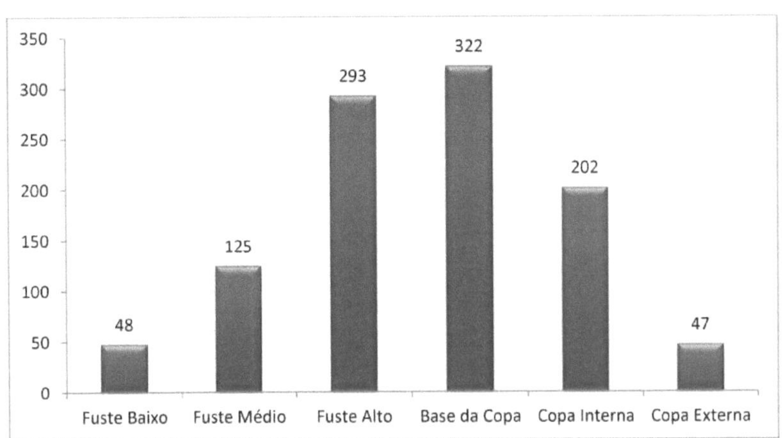

Figura 50: Distribution of the abundance of epiphytic vascular species among the forophytic strata at Sitio Réplica III in the Upstream Area of the Sorocaba/Médio Tietê Hydrographic Basin.

The analysis applied to the distribution of vascular epiphytes in the forophytic strata revealed that the low stem and the outer canopy did not differ from each other, but were significantly different from all the other extracts (Table 48). In general, the high stem and the base of the canopy showed the highest abundances (Figure 50), being significantly different (both) from the middle stem, and in the case of the base of the canopy also different from the inner canopy. This result is similar to that observed at Sitio Réplica II in the same area and also at sites under anthropogenic influence in the Central and Downstream Areas.

The fact that anthropogenic influence alters the development of the epiphytic community is corroborated by the reduction in the number of individuals in the low stem. Although the intermediate strata are more conducive to vascular epiphytism, the high concentration of abundances in these strata plus the dominance of a few species (Table 47) is typical of areas under strong human influence. This observation is pertinent because even in drier forests, as in the case of Sitio Réplica III in the Downstream Area, when the forest fragment is conserved, there is no strong dominance of species and the low stem tends to have a greater number of individuals. The highest number of individuals in the low stem also occurred in Sites Core and Replica I of the Upstream Area, which are well-conserved forest fragments with a similar phytophysiognomy to that found in Site Replica III. This distribution pattern can be used as an indication of the state of conservation of forest remnants.

Table 48: Analysis of the distribution of the richness and abundance of vascular epiphytes among the forophytic strata at Site Replica III of the Upland Area in the Sorocaba/Médio

Tietê hydrographic basin.

^^^-^^^Rich Abundance"----^	Beam Bass	Beam Medium	Beam High	World base	Cup Internal	Cup External
Low stem		**0,00831**	**0,00149**	**5,62E-05**	**0,03444**	0,1194
Medium stem	**0,00727**		0,5647	0,1194	0,5872	**5,62E-05**
High stem	**5,81E-04**	**0,008**		0,3231	0,261	**6,21E-06**
World Cup base	**8,20E-05**	**0,0051**	0,8885		**0,03728**	**1,24E-07**
Internal Cup	**0,04793**	0,2404	0,05936	**0,005899**		**3,91E-04**
External Cup	0,1156	**6,84E-05**	**4,86E-06**	**2,63E-07**	**6,28E-04**	

The variation in richness in the strata, being significantly different (and lower) in both the outer canopy (only four species) and the low stem (nine species) from all the other strata except each other, is indicative of the limitations of the epiphytic community to these strata when the environment is degraded. Although the variation in richness observed in the outer canopy is characteristic of the extreme conditions (especially for water acquisition) faced by the epiphytic community in this stratum, which naturally reduces the number of species adapted to this region of the forophyte. However, the lack of a significant difference between the richness occurring in the outer canopy and the low stem is a reflection of the small number of species recorded in these extremes of the host trees, but it may also be an indication that species occurring in the canopy show great plasticity and, depending on the light conditions, can occur even in the lower stratum of the forest.

Analysis of the vascular epiphytes of the Qualitative Sites (QI, QII, QIII) in the Upstream Area of the Sorocaba/Médio Tietê Hydrographic Basin

The survey of vascular epiphytes in Qualitative Sites QI, QII and QIII in the Upstream Area of the Sorocaba/Médio Tietê river basin was carried out in three different areas. However, it is necessary to make a reservation regarding the use of the mapping carried out by the Forestry Institute of São Paulo, especially the project and image bank of the Forestry Information System of the State of São Paulo. This survey points to large areas of native dense ombrophilous forest vegetation, but the "ground truth" revealed that most of these areas are eucalyptus plantations.

The mapping error, although it seems impossible, can be justified because these eucalyptus plantations really do look like native vegetation, and only at a short distance from these areas is it possible to be sure of their real vegetation composition. Another caveat is that in this area the forests are almost always classified as Dense Ombrophilous Forest, when, in reality, their visible appearance and floristics (considering the vascular epiphytes studied here) is that of an ecotone between Semideciduous Seasonal Forest and Dense Ombrophilous Forest. In addition, it is possible to assume that this mapping error, based on satellite images, could reduce the percentage of native vegetation, which according to CBH-SMT and FABH-SMT (2008) is 12.09%, in the Sorocaba/Médio Tietê basin, since this upstream part of the basin is the one with the highest percentage of vegetation.

The Qualitative Site I (QI) was set up on an area of approximately 45 ha, located between UTM coordinates 272.930 and 7.416.402 in zone 23 South, in the municipality of Itu - SP. The vegetation can be characterized as an ecotone region between the Semideciduous Seasonal Forest and the Dense Ombrophilous Forest. The forest remnant of Qualitative Site II (QII) is approximately 93 ha in size and is located between the UTM coordinates 270.818 and 7.423.666 in zone 23 South, and is characterized as a stretch of Dense Ombrophilous Forest close to the SP 312 highway. Qualitative Site III (QIII), located between UTM coordinates S 282.283 and 7.415.522 in zone 23 South, is a 511 ha forest fragment located in the municipality of Cabreuva-SP, and can be characterized as an area of Dense Ombrophilous Forest. In all three qualitative sites, anthropic influence is visible, such as the presence of trails, traces of fishing equipment near the waterbeds, traps and jiraus (Figure 51), for the illegal capture of wild animals. The areas surrounding the qualitative sites can be characterized by anthropic use, such as agricultural cultivation (especially forestry) and cattle breeding.

Figura 51:   Jirau-type trap present in Qualitative Site I of the Upstream Area.

In the three Qualitative Sites (QI, QII, QIII) in the Upstream Area of the Sorocaba/Médio Tietê basin, 49 species were recorded, belonging to 25 genera and nine families (Table 49).

Table 49 - List of vascular epiphyte species found at the Qualitative Sites in the Montante area of the Sorocaba Mèdio Tietê river basin and their respective Ecological Categories (EC) - HLC: Characteristic holoepiphyte; HLF: Facultative holoepiphyte; HLA: Accidental holoepiphyte; HMP: Primary hemiepiphyte: HMS: Secondary hemiepiphyte. Dispersal form (Disp.) - Zo: Zoochoric; An: Anemochoric. Q: Qualitative Survey Site (1, 2, 3).

153

N	Family	Species	EC	Disp.	Sitio
1	Araceae	Anthurium sellowianum Kunth	HLF	Zo	Q3;
2		Philodendron bipinnatifidum Schott	HMP	Zo	Q3;
	Aspleniaceae				
3	Asplenium scandicinum Kaulf.		HLC	An	Q1;
	Bromeliaceae				
4	Acanthostachys strobilacea (Schult. f.) Klotzsch		HLC	Zo	Q2; Q3;
5		Aechmea bromeliifolia (Rudge) Baker	HLC	Zo	Q2;
6		Aechmea distichantha Lem.	HLC	Zo	Q2;
7		Aechmea nudicaulis (L.) Griseb.	HLC	Zo	Q3;
8		Aechmea sp.	HLC	Zo	Q3;
9		Billbergia amoena (Lodd.) Lindl.	HLC	Zo	Q3;
10		Billbergia distachya (Vell.) Mez	HLC	Zo	Q3;
11		Tillandsia dura Baker	HLC	An	Q3;
12		Tillandsia fasciculata Sw.	HLC	An	Q2;
13		Tillandsia geminiflora Brongn.	HLC	An	Q3;
14		Tillandsia pohliana Mez	HLC	An	Q2;
15		Tillandsia recurvata (L.) L.	HLC	An	Q1; Q2;
16		*Tillandsia stricta* Sol. ex Sims	HLC	An	Q1; Q2;

Keep going...

Table 49 - Continuation...

N	Family Species	EC	Disp.	Sitio
	Bromeliaceae			
17	*Tillandsia tricholepis* Baker	HLC	An	Q1; Q2; Q3;
18	*Tillandsia usneoides* (L.) L.	HLC	An	Q2; Q3;
19	*Vriesea bituminosa* Wawra	HLC	An	Q3;
20	*Vriesea carinata* Wawra	HLC	An	Q3;
21	*Vriesea procera* (Mart. ex Schult. & Schult.f.) Wittm.	HLC	An	Q3;
	Cactaceae			
22	*Cereus alacriportanus* Pfeiff.	HLA	Zo	Q2;
23	*Epiphyllum phyllanthus* (L.) Haw.	HLC	Zo	Q2; Q3;
24	*Lepismium cruciforme* (Vell.) Miq.	HLC	Zo	Q2;
25	*Rhipsalis baccifera* (J.S. Muell.) Stearn	HLC	Zo	Q1;
26	*Rhipsalis campos-portoana* Loefgr.	HLC	Zo	Q2;
27	*Rhipsalis floccosa* Salm-Dyck ex Pfeiff.	HLC	Zo	Q3;
28	*Rhipsalis paradoxa* (Salm-Dyck ex Pfeiff.) Salm-Dyck	HLC	Zo	Q2; Q3;
29	*Rhipsalis teres* (Vell.) Steud.	HLC	Zo	Q1; Q3;
30	*Rhipsalis trigona* Pfeiff.	HLC	Zo	Q1; Q2; Q3;
31	Commelinaceae	HLA	Zo	Q2;

	Tradescantia albiflora Kunth			
	Orchidaceae			
32	*Brasiliorchis gracilis* (Lindl.) R.B. Singer *et al.*	HLC	An	Q3;
33	*Bulbophyllum napellii* Lindl.	HLC	An	Q3;
34	*Catasetum* sp.	HLC	An	Q2;
35	*Cattleya* sp.	HLC	An	Q2; Q3;
36	*Lophiaris pumila* (Lindl.) Braem	HLC	An	Q1;
	Piperaceae			
37	*Peperomia castelosensis* Yunck.	HLC	Zo	Q2;
38	*Peperomia pereskiifolia* (Jacq.) Kunth	HLC	Zo	Q2;
	Polypodiaceae			
39	*Campyloneurum nitidum* (Kaulf.) C. Presl	HLC	An	Q3;
40	*Microgramma crispata* (Fée) R.M.Tryon & A.F.Tryon	HLC	An	Q3;
41	*Microgramma lycopodioides* (L.) Copel.	HLC	An	Q1;
42	*Microgramma squamulosa* (Kaulf.) de la Sota	HLC	An	Q1; Q2; Q3;
43	*Microgramma tecta* (Kaulf.) Alston	HLC	An	Q3;
44	*Pecluma filicula* (Kaulf.) M.G. Price	HLC	An	Q1; Q3;
45	*Pleopeltis hirsutissima* (Raddi) de la Sota	HLC	An	Q1; Q3;
46	*Pleopeltis pleopeltifolia* (Raddi) Alston	HLC	An	Q1; Q3;
47	*Pleopeltis squalida* (Vell.) de la Sota	HLC	An	Q1; Q2;
48	*Serpocaulon latipes* (Langsd. & Fisch.) A.R. Sm	HLC	An	Q1; Q3;
	Pteridaceae			
49	*Vittaria lineata* (L.) Sm.	HLC	An	Q3;

Only 15 species belonging to eight genera and five families were recorded at Sitio QI. Characteristic holoepiphytes were dominant at this site (all 15 species), with no accidental holoepiphytes, facultative holoepiphytes, primary hemiepiphytes or secondary hemiepiphytes recorded. The Polypodiaceae family was the richest at this site with seven species, followed by Bromeliaceae and Cactaceae with three species each. Aspleniaceae and Orchidaceae had one species each.

In Site QII of the Montante Area, with 22 species belonging to 13 genera and six families, characteristic holoepiphytes accounted for 20 species, followed by accidental holoepiphytes and facultative holoepiphytes, both with one species. The Bromeliaceae family with nine species and Cactaceae with six species were dominant at this site. Polypodiaceae, Orchidaceae and Piperaceae accounted for two species each and the Commelinaceae family had one species.

Site QIII was the most diverse of all the sites in the Montante Area, with 31 species, 18 genera and six families of vascular epiphytes. In this site, the characteristic holoepiphytes were dominant, accounting for 28 species, followed by facultative holoepiphytes with two species and primary hemiepiphytes with one species. No accidental holoepiphytes or secondary hemiepiphytes were recorded. Bromeliaceae (Figure 52) with 12 species and Polypodiaceae, with eight, were dominant at this site. The Cactaceae family had five species, Orchidaceae three, Araceae two and Pteridaceae one.

Figura 52: Acanthostachys strobilacea (Schult. f.) Klotzsch (Bromeliaceae), present in two of the three qualitative sites in the Montante Area: A - epiphytic habit; B - fruit.

The richness of vascular epiphytes at this site can be considered low when compared to studies carried out in Dense Ombrophilous Forest by Blum (2010) - 277 species; Petean (2009) - 159 species; Kersten (2006) - 349 species; Breier (2005) - 161 species; Petean (2003) - 97 species; Schütz-Gatti (2000) - 175 species; Fontoura et al. (1997) - 293 species; and Hertel (1950) - 101 species, and also lower than that observed in Semideciduous Seasonal Forest by Rogalski and Zanin (2003) - 70 species; Giongo and Waechter (2004) - 57 species and Cervi and Borgo (2007) - 56 species. However, it is similar to the results obtained by Dislich and Mantovani (1998) - 34 species, Bonnet et al. (2011) - 35 species (average of six collection seasons), by Borgo et al. (2002) - 32 species, Breier (2005) - 25 species, Dettke et al. (2008) - 29 species and Bataghin et al. (2010) - 21 species.

It is important to mention that Bonnet et al. (2011), studying several fragments of Semideciduous Seasonal Forest in the basin of the Tibagi River - PR (a similar study to the one carried out in the present study), revealed an average of 35 species per forest area, a similar number to the one found here. One factor that may have negatively influenced the diversity of this site is anthropic interference in the environment. The obvious access of people to the Qualitative Sites of the Upland Area may be responsible for the removal of epiphytic species, especially Bromeliaceae and Orchidaceae - which are of ornamental interest, reducing the diversity of this species in the sites studied. The anthropogenic disturbance factor, especially the removal of species, seems to be accentuated in forest fragments close to large urban centers or where there is a greater demand on the natural environment for human survival.

Several studies have reported that changes in the landscape negatively influence the diversity and abundance of vascular epiphytes (ENGWALD et al., 2000; BONNET; QUEIROZ, 2000; BARTHLOTT et al., 2001; BATAGHIN et al., 2008; DETTKE et al.., 2008), however, the conservation of forest remnants, especially those not protected by UCs, should not be overlooked, as the forest fragmentation of the Sorocaba/Médio Tietê basin may have isolated or even restricted the epiphytic community in these forest fragments.

The richest families in the Upstream Area Qualitative Sites (QI, QII, QIII) were Bromeliaceae with 18 species, Polypodiaceae with 10 species and Cactaceae with nine species. The Orchidaceae family had five species. Araceae and Piperaceae had two species each. Aspleniaceae, Commelinaceae and Pteridaceae had one species each. In general, characteristic holoepiphytes were dominant at the sites, accounting for 44 species (90%), followed by facultative holoepiphytes with two species (4%), and accidental holoepiphytes, also with two species (4%). Secondary hemiepiphytes appeared with one species each. No primary hemiepiphytes were recorded at the qualitative sites in the Upstream Area.

Of the total number of species, 28 (56%) showed anemochoric dispersal and 21 (44%) zoochoric dispersal. This result partly reflects that observed by Gentry and Dodson (1987a), who recorded 2/3 of the species as anemochoric. However, the reduction in the proportion of anemochoric/zoochoric species provides an indication that small anthropically altered forest fragments or those with low structural complexity (observed in the qualitative sites in this area) tend to increase the number of epiphytic species with a zoochoric dispersal syndrome.

Analysis of vascular epiphytes in the Ecotone between Semideciduous Seasonal Forest and Dense Ombrophilous Forest in the Upland Area of the Sorocaba/Médio Tietê Watershed

The analysis of the distribution of vascular epiphyte abundances between the sites showed no significant differences between the Core Site and its replicas. The values obtained were as follows: the analysis between the Core Site and Replica Site I showed t = 1.470 and p = 0.072; between the Core Sites and Replica II, t = 1.679 and p = 0.057; and between the Core Sites and Replica III, t = 1.493 and p = 0.068. The absence of a significant difference between the abundances of the Core Sites and their replicas may indicate that the vascular epiphytic community that develops in the Semideciduous Seasonal ForestZDense Ombrophilous Forest Ecotone of the Upstream Area in the watershed can be considered similar in terms of the number of individuals in the different sites studied.

The same analysis applied to the presence/absence of species showed the Core Site to be different from all its replicas, with the following results: the analysis between Sitios Core and Replica I showed t = 6.496 and p < 0.0001; between Sitio Core and Sitio Réplica II, t = 6.796 and p < 0.0001; between Sitios Core and Réplica III, t = 9.043 and p < 0.0001; between Core and Q I, t = 13.521 and p < 0.0001; between Core and Q II, t = 10.193 and p < 0.0001; and between Core and Q III, t = 7.825 and p < 0.0001. This reflects the diversity of vascular epiphytic species found in the Montante Area as a whole and the importance of conserving forest areas, especially if these fragments are protected by Conservation Units, although forests in private areas contribute an important part to this vascular epiphytic diversity. Another important observation is that these results provide an indication that there is a different partitioning of niches within the forest in the Upland Area, i.e. a different number of individuals of different species occupy more or less space on the forophytes, depending on the site studied, although the number of individuals at these sites tends to be very similar.

The Jaccard Similarity Index, the results of which can be seen in Figure 53, showed similarities of: 13.7% between Sitio Core and Sitio Réplica I; 10.7% between Sitio Core and Sitio Réplica II; 15.1% between Sitio Core and Sitio Réplica III; 4.44% between Sitio Core and Sitio Q I; 5.2% between Sitio Core and Sitio Q II; and 11.11% between Sitio Core and Sitio Q III. The greatest similarities in this area of the watershed were observed between Replica Site I and Replica Site II (35.8%) and between Replica Site II and Replica Site III (31.3%). This low level of similarity between the Core Site and the Replica Sites corroborates the idea that the vascular epiphytic community found in the Upstream Area of the Sorocaba/Médio Tietê basin is heterogeneous, revealing the great importance of the Conservation Unit studied in the composition of the vascular epiphytic flora, but highlighting the fact that forest fragments not protected by Conservation Units are responsible for almost 35% of the species found in this area of the basin, which underscores their importance.

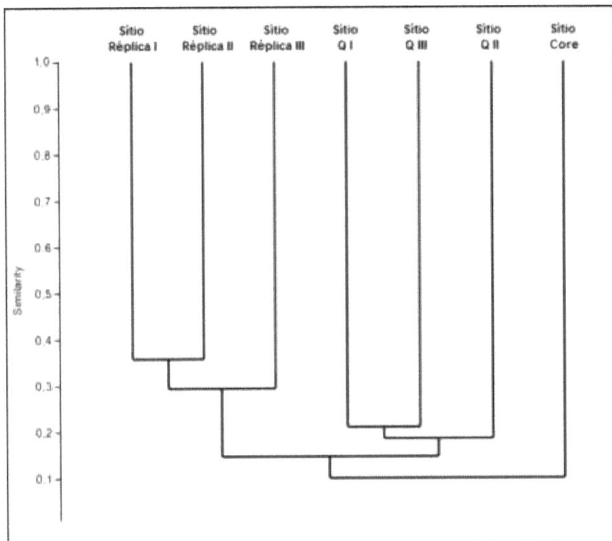

Figura 53: Dendrogram (UPGMA) of Jaccard's similarity between sites in the Ecotone between Semideciduous Seasonal Forest and Dense Ombrophilous Forest in the upstream area of the Sorocaba/Médio Tietê basin.

Another interesting aspect is the similarity between the vascular epiphytes sampled in the Montante Area of the Sorocaba/Médio Tietê basin and other studies carried out on vascular epiphytes in different forest formations in Brazil, both in areas of Semideciduous Seasonal Forest and in areas of Dense Ombrophilous Forest (Table 50).

Table 50 - Jaccard's similarity index (J) between the vascular epiphytic species from the Montante Area of the Sorocaba/Médio Tietê basin and other studies of vascular epiphytes carried out in Semideciduous Seasonal Forest and Dense Ombrophilous Forest in Brazil. FES: Semideciduous Seasonal Forest; FOD: Dense Ombrophilous Forest; FOM: Mixed Ombrophilous Forest.

Source	Location (Phytophysiognomy)	Species	Families	J
Aguiar et al. 1981	Montenegro - RS (FES)	17	5	0,084
Dislich and Mantovani 1998	Sao Paulo - SP (FES)	34	9	0,191
Borgo et al. 2002	Fenix - PR (FES)	32	10	0,109
Rogalski and Zanin 2003	Marcelino Ramos - RS (FES)	70	8	0,099
Giongo and Waechter 2004	Eldorado do Sul - RS (FES)	57	13	0,147
Breier 2005	Assis - SP (FES)	25	9	0,089
Cervi and Borgo 2007	Foz do Iguaçù - PR (FES)	56	13	0,092
Dettke et al. 2008	Maringa - PR (FES)	29	8	0,091
Bataghin et al. 2010	Iperó - SP (FES)	21	5	0,119

Bonnet et al. 2011	Tibagi River Basin - PR (FES)	60	13	0,259
Geraldino et al. 2010	Campo Mourao - PR (FES/FOM)	61	12	0,164
Bonnet et al. 2011	BH Tibagi - PR (FES/FOM)	188	24	0,251
Menini-Neto et al. 2011	Descoberto - MG (FES)	59	10	0,122
Breier 2005	Sete Barras - SP (FOD)	161	27	0,293
Zipparro et al. 2005	Intervales/Base Saibadela - SP (FOD)	55	10	0,101
Kersten 2006	Piraquara - PR (FOD/FOM)	143	27	0,207
Petean 2009	Antonina - PR (FOD)	159	23	0,288
Blum 2010	Morretes - PR (FOD)	277	30	0,283
Menini-Neto et al. 2011	Lima Duarte/Sta R. de Ibitipoca - MG (FOD)	113	9	0,091

The low level of similarity in relation to the majority of studies seen in Table 50 may be associated with the great sensitivity of the epiphytic community to climatic and microclimatic variations (BENZING, 1990; 1995), given the environmental changes suffered by the forests in the Sorocaba/Médio Tietê basin (ZERO REPORT, 2005), or with the particular climatic and microclimatic characteristics of the sites analyzed. Environmental changes, especially the fragmentation and loss of forests, are the main factors responsible for the alterations and

microclimatic variations in forests. Dettke et al. (2008) and Barthlott et al. (2001) point out that the occurrence of epiphytic species is related to the integrity of the forest and, consequently, to the existence of favorable climatic conditions for them. However, the variations caused by the natural distribution of species in different forest phytophysiognomies and, above all, the occurrence of different species according to latitudinal variations (WAECHTER, 1998) is one possibility to explain the low similarity between the epiphytic communities that occur in the area studied and the species observed by different authors cited in Table 50.

The irregular geographical distribution pointed out by Kersten (2006) may be another important factor behind the low similarity observed above. It is interesting to note that the greatest similarities occur between similar forest formations that are relatively close together on a landscape scale, while the lowest similarity was observed in relation to an area that is geographically distant. The area that showed the greatest similarity was Sete Barras - SP (BREIER, 2005), in which Dense Ombrophilous Forest predominates and which is one of the closest. Although there is a general tendency for epiphytic diversity to decrease from tropical regions towards the poles (SMITH, 1962), the dependence on humidity, absorbed directly from the air, makes moist forests centers of epiphytic biodiversity (BENZING, 1990; SCHÜTZ-GATTI, 2000; KERSTEN; SILVA, 2001). This may contribute to the greater diversity of the core site compared to the replicate sites which, in general, are drier forest formations and which, because they are not protected as a Conservation Unit, tend to be more susceptible to anthropogenic influences.

The quantitative assessment of the vascular epiphytes from all the sites (core and its replicates) in the Upstream Area is shown in Table 51. The Upstream Area of the Sorocaba/Médio Tietê basin has climatic characteristics that are different from those observed in the other areas (Central and Downstream) of the river basin, especially with regard to the distribution of moisture during the year - which is well distributed throughout most of the year. In the Upstream Area, there are mainly species that are resistant to periods of water deficit, such as some genera of the Polypodiaceae and Bromeliaceae families, especially the Tillandsia and Pleopeltis genera, However, there are also species that are characteristic of humid forests, such as Araceae, Orchidaceae, Dryopteridaceae and genera of Bromeliaceae, such as *Vriesea,* which are responsible for more than 30% of the epiphytic value of importance (VIE) in the Upland Area.

Table 51 - Vascular epiphytes of the Upstream Area of the Sorocaba/Médio Tietê Hydrographic Basin (Semideciduous Seasonal Forest/Dense Ombrophilous Forest Ecotone), classified according to the value of epiphytic importance - nr: absolute number of occurrences in the strata; far: absolute frequency in the strata; ni: absolute number of occurrences in the forophytic individuals; fai: absolute frequency in the forophytic individuals; vt (total value): sum of the abundance estimates; vie: epiphytic importance value; nota: average score obtained.

Species	nr	far	ni	fai	vt	vie	note
Microgramma squamulosa	537	17,0	162	45,0	900	17,77	1,68
Pleopeltis pleopeltifolia	267	8,4	101	28,1	388	7,66	1,45
Pleopeltis hirsutissima	251	7,9	105	29,2	379	7,48	1,51
Vriesea incurvata	207	6,5	67	18,6	342	6,75	1,65
Campyloneurum acrocarpon	117	3,7	56	15,6	169	3,34	1,44
Dichaea trulla	103	3,3	48	13,3	163	3,22	1,58
Gomesa recurva	83	2,6	36	10,0	147	2,90	1,77
Serpocaulon latipes	84	2,7	40	11,1	132	2,61	1,57
Tillandsia geminiflora	95	3,0	36	10,0	130	2,57	1,37
Billbergia distachya	69	2,2	36	10,0	114	2,25	1,65
Anthurium sellowianum	60	1,9	23	6,4	113	2,23	1,88
Philodendron propinquum	67	2,1	23	6,4	109	2,15	1,63
Campyloneurum nitidum	55	1,7	30	8,3	91	1,80	1,65
Serpocaulon catharinae	53	1,7	25	6,9	83	1,64	1,57
Vriesea altodaserrae	43	1,4	20	5,6	81	1,60	1,88

Species	nr	far	ni	fai	vt	vie	note
Elaphoglossum ornatum	56	1,8	30	8,3	75	1,48	1,34
Pecluma truncorum	43	1,4	17	4,7	69	1,36	1,60
Vriesea carinata	47	1,5	19	5,3	68	1,34	1,45
Sinningia douglasii	35	1,1	13	3,6	58	1,15	1,66
Begonia fruticosa	35	1,1	12	3,3	53	1,05	1,51
Vittaria lineata	34	1,1	15	4,2	49	0,97	1,44
Peperomia urocarpa	28	0,9	19	5,3	49	0,97	1,75
Campyloneurum repens	27	0,9	13	3,6	45	0,89	1,67
Microgramma persicariifolia	24	0,8	9	2,5	45	0,89	1,88
Marcgravia polyantha	25	0,8	9	2,5	42	0,83	1,68
Codonanthe gracilis	20	0,6	5	1,4	42	0,83	2,10
Microgramma percussa	28	0,9	11	3,1	41	0,81	1,46
Vriesea gigantea	19	0,6	13	3,6	41	0,81	2,16
Microgramma lycopodioides	23	0,7	8	2,2	39	0,77	1,70
Varicose coppensia	22	0,7	11	3,1	39	0,77	1,77
Polytaenium cajenense	28	0,9	12	3,3	38	0,75	1,36
Pecluma sp.	22	0,7	10	2,8	37	0,73	1,68
Serpocaulon fraxinifolium	22	0,7	9	2,5	36	0,71	1,64
Neoregelia laevis	16	0,5	9	2,5	36	0,71	2,25
Lophiaris pumila	22	0,7	14	3,9	34	0,67	1,55
Microgramma vacciniifolia	21	0,7	8	2,2	34	0,67	1,62
Vriesea rodigasiana	22	0,7	11	3,1	33	0,65	1,50

Keep going...

Table 51 - Continuation...

Species	nr	far	ni	fai	vt	vie	note
Elaphoglossum glabellum	*17*	0,5	8	2,2	30	0,59	1,76
Aechmea bromeliifolia	*15*	0,5	7	1,9	30	0,59	2,00
Asplenium scandicinum	*18*	0,6	9	2,5	28	0,55	1,56
Campylocentrum aromaticum	*21*	0,7	12	3,3	27	0,53	1,29
Aechmea distichantha	*15*	0,5	9	2,5	26	0,51	1,73
Notylia cf. *longispicata*	*21*	0,7	12	3,3	24	0,47	1,14
Saundersia mirabilis	*17*	0,5	9	2,5	24	0,47	1,41
Billbergia zebrina	*15*	0,5	10	2,8	24	0,47	1,60
Niphidium crassifolium	*13*	0,4	7	1,9	22	0,43	1,69
Tillandsia recurvata	*13*	0,4	5	1,4	22	0,43	1,69
Canistrum lindenii	*10*	0,3	8	2,2	21	0,41	2,10
Asplenium mucronatum	*12*	0,4	4	1,1	19	0,38	1,58
Rhipsalis teres	*11*	0,3	5	1,4	19	0,38	1,73

Species	nr	far	ni	fai	vt	vie	note
Miltonia sp.	*11*	0,3	7	1,9	17	0,34	1,55
Campylocentrum cf. *grisebachii*	*13*	0,4	7	1,9	16	0,32	1,23
Serpocaulon sehnemii	*11*	0,3	7	1,9	16	0,32	1,45
Anthurium acutum	*11*	0,3	6	1,7	15	0,30	1,36
Tillandsia sp.	*11*	0,3	6	1,7	15	0,30	1,36
Gomesa glaziovii	*10*	0,3	5	1,4	15	0,30	1,50
Vriesea hieroglyphica	*7*	0,2	4	1,1	15	0,30	2,14
Pleopeltis macrocarpa	*9*	0,3	4	1,1	14	0,28	1,56
Asplenium sp.	*8*	0,3	3	0,8	14	0,28	1,75
Scaphyglottis modesta	*7*	0,2	4	1,1	13	0,26	1,86
Blechnum binervatum	*8*	0,3	4	1,1	12	0,24	1,50
Brasilidium sp.	*7*	0,2	5	1,4	12	0,24	1,71
Tillandsia araujei	*8*	0,3	4	1,1	11	0,22	1,38
Elaphoglossum lingua	*6*	0,2	5	1,4	11	0,22	1,83
Pleopeltis squalida	*6*	0,2	2	0,6	11	0,22	1,83
Tillandsia dura	*7*	0,2	4	1,1	10	0,20	1,43
Peperomia castelosensis	*6*	0,2	3	0,8	10	0,20	1,67
Epidendrum ansiferum	*5*	0,2	3	0,8	10	0,20	2,00
Philodendron vargealtense	*5*	0,2	2	0,6	10	0,20	2,00
Dichaea pendula	*7*	0,2	4	1,1	9	0,18	1,29
Anthurium longifolium	*5*	0,2	4	1,1	9	0,18	1,80
Peperomia catharinae	*4*	0,1	2	0,6	9	0,18	2,25
Peperomia alata	*4*	0,1	2	0,6	7	0,14	1,75
Philodendron appendiculatum	*4*	0,1	2	0,6	7	0,14	1,75
Vriesea platynema	*4*	0,1	3	0,8	7	0,14	1,75
Asplenium auritum	*4*	0,1	2	0,6	6	0,12	1,50
Grobya galeata	*3*	0,1	2	0,6	6	0,12	2,00
Asplenium pteropus	*5*	0,2	4	1,1	5	0,10	1,00
Cyclopogon multiflorus	*4*	0,1	2	0,6	5	0,10	1,25
Peperomia trineura	*3*	0,1	1	0,3	5	0,10	1,67

Keep going...

Table 51 - Continuation...

Species	nr	far	ni	fai	vt	vie	note
Prosthechea glumacea	3	0,1	2	0,6	5	0,10	1,67
Rhipsalis trigona	3	0,1	1	0,3	5	0,10	1,67
Lepismium lumbricoides	2	0,1	1	0,3	5	0,10	2,50
Nidularium rutilans	2	0,1	1	0,3	5	0,10	2,50
Peperomia glabella	2	0,1	1	0,3	5	0,10	2,50

Gomesa sp.	3	0,1	2	0,6	4	0,08	1,33
Philodendron corcovadense	3	0,1	1	0,3	4	0,08	1,33
Catasetum atratum	2	0,1	2	0,6	4	0,08	2,00
Catasetum fimbriatum	2	0,1	2	0,6	4	0,08	2,00
Octomeria grandiflora	2	0,1	2	0,6	4	0,08	2,00
Capanemia micromera	3	0,1	2	0,6	3	0,06	1,00
Polybotrya cylindrica	3	0,1	2	0,6	3	0,06	1,00
Elaphoglossum glaziovii	2	0,1	1	0,3	3	0,06	1,50
Encyclia patens	2	0,1	1	0,3	3	0,06	1,50
Prosthechea cf. *bulbosa*	2	0,1	1	0,3	3	0,06	1,50
Rodriguezia decorates	2	0,1	1	0,3	3	0,06	1,50
Stelis deregularis	2	0,1	1	0,3	3	0,06	1,50
Tillandsia tenuifolia	2	0,1	1	0,3	3	0,06	1,50
Ceradenia albidula	2	0,1	1	0,3	2	0,04	1,00
Aechmea nudicaulis	1	0,0	1	0,3	2	0,04	2,00
Nematanthus striatus	1	0,0	1	0,3	2	0,04	2,00
Polystachya concreta	1	0,0	1	0,3	2	0,04	2,00
Rhipsalis campos-portoana	1	0,0	1	0,3	2	0,04	2,00
Stigmatopteris caudata	1	0,0	1	0,3	2	0,04	2,00
Polystachya estrellensis	1	0,0	1	0,3	1	0,02	1,00

The species with the highest importance value for the Montante Area was *Microgramma squamulosa* (Polypodiaceae) with an epiphytic importance value (EVI) of 17.77 and an average score of 1.68; this species occurred in 45% of the forophytes and 17% of the strata sampled. *Pleopeltis pleopeltifolia* (Polypodiaceae), with a VIE of 7.66 and an average score of 1.45, occurring in 28.1% of the forophytes and 8.4% of the strata, was the second most important species in the Montante Area. *Pleopeltis hirsutissima* (Polypodiaceae) had a VIE of 7.48 and an average score of 1.51, and was found in 29.2% of the forophytes and 7.9% of the strata. In addition to these Polypodiaceae, *Vriesea incurvata* (Bromeliaceae) had a VIE of 6.75 and an average score of 1.65, occurring in 18.6% of the forophytes and 6.5% of the strata. *Campyloneurum acrocarpon* (Polypodiaceae) had a VIE of 3.34 and an average score of 1.44, while *Dichaea trulla* (Orchidaceae) had a VIE of 3.22 and an average score of 1.58 (Figure 54). These six species were responsible for more than 45% of the epiphytic importance value in the Ecotone between Semideciduous Seasonal Forest and Dense Ombrophilous Forest in the Upstream Area of the Sorocaba/Médio Tietê basin.

Figura 54: Dichaea trulla Rchb. f. (Orchidaceae), the species with the highest epiphytic importance value among the orchids recorded in the Upland Area.

The Polypodiaceae, Bromeliaceae and Orchidaceae families are often among the most common in Brazilian studies, both in Seasonal Semideciduous Forest (DISLICH; MANTOVANI, 1998; ROGALSKI; ZANIN, 2003; GIONGO; WAECHTER, 2004; BREIER, 2005; DETTKE et al, 2008; BATAGHIN et al., 2010), as well as in areas of dense ombrophilous forest (BREIER, 2005; KERSTEN, 2006; PETEAN, 2009; BLUM, 2010). The Polypodiaceae family was the most important in the area with a VIE of 50.41, followed by Bromeliaceae with a VIE of 20.46% and species from the Orchidaceae family, responsible for 11.75% of the area's VIE. Although the upstream area of the Sorocaba/Médio Tietê basin is wetter than the central and downstream areas, the success of species from the Polypodiaceae family, which tend to be more resistant to water deficit and/or temperature variation, may be indicative of the type of forest environment in this area of the basin.

There was a significant difference between the shape of the vertical distribution of vascular epiphytes (abundance) occurring in the Core Site of the Montante Area and its three replicas ($p < 0.05$). The comparison of the strata of the Core Site and the equivalent strata in Replica Sites I, II and III revealed that the vertical distribution of vascular epiphytes is significantly different in all strata, except in the paired comparisons between the Core Site and its replicas in the mid-stem of Replica Site I and in the Outer Cup of Replica Sites II and III.

In the distribution of epiphytes in the strata, looking at all the sites in the Montante Area, the base of the canopy stood out as the stratum with the highest epiphytic abundance (Figure 55), with an abundance value (VA) of 1626, followed by the high stem with VA = 1257, the inner canopy with VA = 975 and the middle stem, low stem and outer canopy with abundance values of 712, 330 and 164, respectively.

165

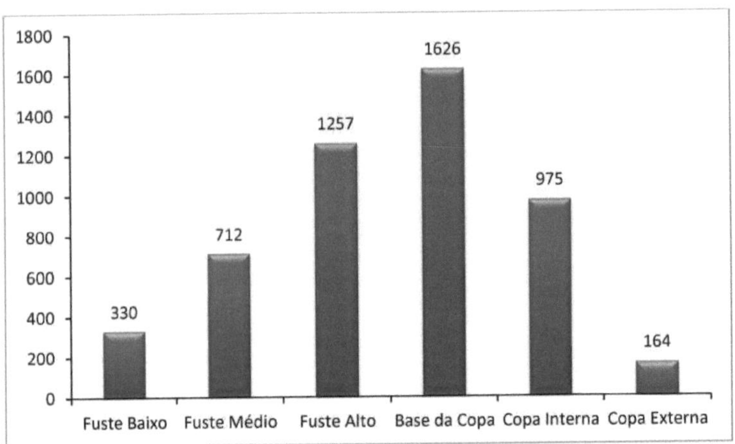

Figura 55:   Distribution of the abundance of vascular epiphytic species among the forophytic strata in the upland area of the Sorocaba/Médio Tietê river basin.

The vertical distribution of the epiphytic community in the upstream area, based on species abundance, showed that there were significant differences between the low stem and all the other forophyte strata (Table 52). Similarly, the outer canopy was significantly different from all the extracts. The high stem and the base of the crown, the most abundant strata in the Upland Area, showed a significant difference from the middle stem. This pattern of vertical distribution of the epiphytic vascular community in the Montante Area is strongly influenced by the records obtained at the Core Site in this area, given the large number of individuals recorded at this site, and also because it presents a very similar model.

Although the variation in the abundance of forophytes was expected, this form of distribution is characteristic of better conserved forest areas, especially as the low stem differs from the outer canopy in terms of abundance. This significant variation between these strata was only observed in one site in the Downstream Area (a conserved forest fragment) and in two sites in the Upstream Area. It is also worth highlighting the importance of climatic conditions and also the type of forest involved (FOD), which provides the best microclimatic conditions for vascular epiphytism.

Table 52: Analysis of the distribution of the richness and abundance of vascular epiphytes among the forophytic strata in the Upstream Area of the Sorocaba/Médio Tietê hydrographic basin.

^^-^^^Richness Abundance^'^^^	Beam Bass	Beam Medium	High stem	World base	Cup Internal Cup	Cup External

Low stem		0,002409	9,23E-07	2,94E-10	0,000184	1,10E-05
Medium stem	0,002		0,051	0,000533	0,4677	6,97E-13
High stem	4,69E-04	0,031		0,1184	0,2186	9,32E-19
World Cup base	3,09E-04	0,009	0,208		0,005684	9,22E-24
Internal Cup	0,019	0,208	0,242	0,086		5,63E-15
External Cup	0,0312	2,93E-04	0,0001	0,0001	0,006	

Several authors have postulated that epiphytes are unevenly distributed throughout forophytes, varying vertically in the number of individuals and species, as well as their specific composition (STEEGE; CORNELISSEN, 1989; BROWN, 1990; WAECHTER, 1992). This was partly observed for the vascular epiphytic community of the Upland Area. The species richness recorded in both the low stem and the outer canopy was significantly different from that recorded in all the other strata (Table 52), as was that recorded at the base of the canopy in relation to the middle stem and the inner canopy.

These variations indicate the importance of the base of the canopy for the establishment and development of the epiphytic community, both because it is home to a greater number of species (89) and because it provides support for most of the individuals (Figure 56). The results observed in the outer canopy (16 spp.) reflect the restricted group of species that are able to survive the great microclimatic fluctuations of this stratum, and are able to overcome the water stress of this region. The species found in the outer canopy generally show great environmental plasticity, with 12 of the 16 occurring in all strata in the upstream area.

In general, the same groups of species recorded at Sitio Core can be observed in this area of the hydrographic basin: a) a more generalist group in which almost all species occur from the canopy to the ground; b) the most demanding group in terms of moisture acquisition and prefer shadier environments, being restricted to the lower strata, usually low and medium stem, sometimes reaching the base of the canopy; and c) a third group which occurs in the intermediate regions of the forophytes, and which need better conditions for anchoring propagules, or are demanding in terms of the presence of suspended soil, which increases humidity even though they have better access to light. This corroborates what was proposed by Benzing (1990), who points out that the vertical evolution of vascular epiphytes occurred (and still occurs) through the exchange of more restrictive spaces for the acquisition of water in exchange for better light conditions. The similarity between the species that occur in the intermediate strata can be seen in Figure 56.

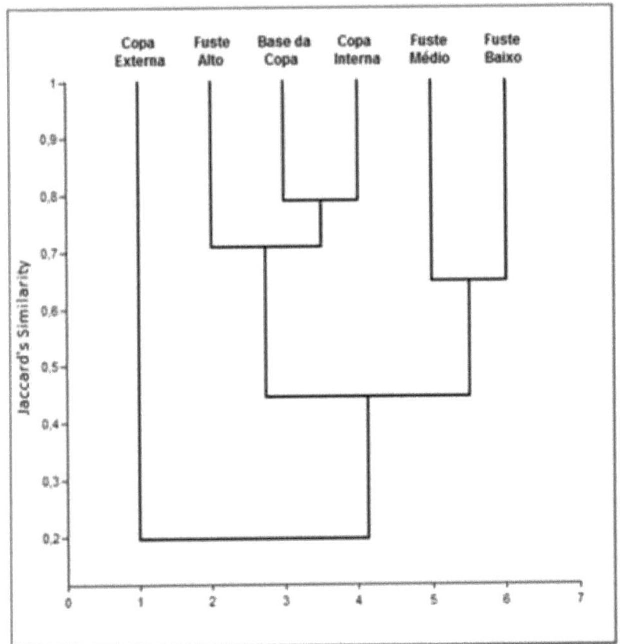

Figura 56: Dendrogram (UPGMA) of Jaccard's similarity between strata in the Semideciduous Seasonal Forest and Dense Ombrophilous Forest ecotone in the upstream area of the Sorocaba/Médio Tietê basin.

The greater similarity between the base of the canopy and the inner canopy may be related to the presence of places that tend to have an accumulation of suspended soil, in the case of the base of the canopy, which increases the retention of nutrients and moisture, or by the presence of branch insertion points, in the case of the inner canopy, increasing the available area which favors the establishment and development of epiphytes (BENZING, 1990; KERSTEN, 2006). The similarities in the distribution of vascular epiphytic community abundances between the two strata mentioned and the low and medium canopies may also be related to the dynamic balance that epiphytes establish in relation to water stress and light availability, factors that have an effect on the distribution of epiphytic species in general. In addition, the outer canopy was the stratum that was least related to the other strata in terms of abundance in the Upstream Area of the basin, an indication that the species in this area are highly sensitive to limiting factors, be they related to reduced humidity or variations in light and temperature during certain periods of the year.

Analysis of the Vascular Epiphytes of the Sorocaba/Médio Tietê River Basin

The floristic survey of vascular epiphytes carried out in these 21 sites (forest fragments) found 176 species (Downstream Area - 56 spp., Central Area - 64 spp. and Upstream Area - 139 spp.), belonging to 66 genera and 14 families (Table 53), of which 50 species are under some degree of threat of extinction. Shannon's diversity index for these sites as a whole was H' = 3.695, equability (J) was 0.713 and Margalef's richness (d) was 18.39.

Table 53 - List of vascular epiphyte species found in the Sorocaba Mèdio Tietê basin and their respective Ecological Categories (EC) HLC: Characteristic Holoepiphyte; HLF: Facultative Holoepiphyte; HLA: Accidental Holoepiphyte; HMP: Primary Hemiepiphyte; HMS: Secondary Hemiepiphyte. Central Area (C), Downstream Area (J) and Upstream Area (M). Disp: Dispersal Syndrome - Zo: Zoochoric, An: Anemochoric. Reg: HUFSCar herbarium registration number (Im: Digital image). In bold, species cited in the Red Book of Brazilian Flora: ↑: Endangered, §: Vulnerable, ∞: Near Threatened, *: Least Concern.

Family	Species	EC	Area	Disp	Reg.
ARACEAE					
1	Anthurium acutum N.E. Br.	HLA	M	Zo	8485
2	Anthurium comtum Schott	HLF	C	Zo	8486
3	Anthurium longifolium **(Hoffm.) G. Don** *	HLC	M	Zo	8499
4	Anthurium sellowianum Kunth	HLF	M	Zo	Im
5	Philodendron appendiculatum Nadruz & S.J. Mayo	HMS	M/J	Zo	8487
6	Philodendron bipinnatifidum Schott	HMP	M/C/J	Zo	Im
7	Philodendron eximium Schott	HMP	C	Zo	8488
8	Philodendron corcovadense Kunth	HMP	M	Zo	8385
9	Philodendron propinquum Schott	HMS	M	Zo	8500
10	Philodendron vargealtense **Sakur.** *	HMP	M	Zo	8394
ASPLENIACEAE					
11	Asplenium auritum Sw.	HLF	M	An	8497
12	Asplenium mucronatum C. Presl	HLA	C	An	8483
13	Asplenium pteropus Kaulf.	HLC	M	An	8481
14	Asplenium pulchellum Raddi	HLC	M	An	8498
15	Asplenium scandicinum Kaulf.	HLC	M	An	8482
16	Asplenium sp.	HLC	M	An	8438
BEGONIACEAE					
17	Begonia fruticosa **A. DC.** *	HMS	M	An	8484
BLECHNACEAE					
18	*Blechnum binervatum* (Poir.) C.V. Morton & Lellinger	HMS	M	An	8501

Table 53 - Continuation...

Family	Species	EC	Area	Disp	Reg.

BROMELIACEAE

19	*Acanthostachys strobilacea* (Schult. f.) Klotzsch	HLC	M/C/J	Zo	8431
20	Aechmea apocalyptica REITZ§	HLF	C/J	Zo	8455
21	Aechmea bromeliifolia **(Rudge) Baker** *	HLC	M/C/J	Zo	8432
22	*Aechmea racinae* L.B. Sm.	HLC	M	Zo	8510
23	Aechmea distichantha **Lem.** *	HLC	M/C/J	Zo	8503
24	Aechmea nudicaulis **(L.) Griseb.** *	HLC	M/C	Zo	8434
25	*Aechmea* sp.	HLC	M	Zo	8456
26	*Billbergia amoena* (Lodd.) Lindl.	HLC	M/C/J	Zo	8379
27	Billbergia distachya **(Vell.) Mez** *	HLC	M/C	Zo	8445
28	*Billbergia porteana* Brongn. ex Beer	HLC	C	Zo	8457
29	Billbergia zebrina **(Herb.) Lindl.** *	HLC	M/C	Zo	8433
30	*Canistropsis billbergioides* (Schult. & Schult.f.) Rudder	HLC	M	Zo	8511
31	*Canistrum lindenii* (Regel) Mez	HLC	M	Zo	Im
32	*Neoregelia laevis* (Mez) L.B. Sm.	HLC	M	Zo	Im
33	*Nidularium rutilans* E. Morren	HLC	M	Zo	8456
34	Nidularium innocentii **Lem.** *	HLC	M	Zo	8505
35	***Tillandsia araujei* Mezf**	HLC	M	An	8377
36	*Tillandsia funckiana* Baker	HLC	C/J	An	8458
37	*Tillandsia dura* Baker	HLC	M	An	8459
38	*Tillandsia fasciculata* Sw.	HLC	M	An	8514
39	***Tillandsia geminiflora* Brongn.** *	HLC	M	In	8427
40	***Tillandsia linearis* Vell.** *	HLC	M	In	8462
41	*Tillandsia pohliana* Mez	HLC	M	In	8463
42	*Tillandsia recurvata* (L.) L.	HLC	M/C/J	In	8429
43	*Tillandsia* sp.	HLC	M/C	In	8425
44	*Tillandsia stricta* Sol. ex Sims	HLC	M/C/J	In	8428
45	***Tillandsia tenuifolia* L.** *	HLC	M	An	8509
46	***Tillandsia tricholepis* Baker** *	HLC	M/C/J	An	8430
47	***Tillandsia usneoides* (L.) L.** *	HLC	M/C/J	An	8460
48	*Vriesea altodaserrae* L.B. Sm.	HLC	M	An	8461
49	***Vriesea bituminosa* Wawra** *	HLC	M/J	In	8515
50	***Vriesea carinata* Wawra** *	HLC	M	An	8513
51	*Vriesea fenestralis* Linden & André	HLC	C	In	8444
52	***Vriesea flammea* L.B. Sm.** *	HLC	M	In	8507
53	*Vriesea friburgensis* Mez	HLF	C	In	8426
54	***Vriesea gigantea* Mart. ex Schult. f.** *	HLC	M	An	8512
55	***Vriesea hieroglyphica* (Carrière) E. Morren** *	HLC	M	An	Im
56	***Vriesea incurvata* Gaudich.** *	HLC	M	An	8506
57	***Vriesea platynema* Gaudich.** *	HLC	M	In	8464
58	***Vriesea procera* (Mart. ex Schult. & Schult.f.) Wittm.** *	HLC	M/C/J	An	8508

Table 53 - Continuation...

Family	Species	EC	Area	Disp	Reg.
BROMELIACEAE					
59	Vriesea rodigasiana **E. Morren** *	HLC	M	An	8504
60	*Vriesea vagans* (L.B. Sm.) L.B. Sm.	HLC	M	An	Im
CACTACEAE					
61	*Cereus alacriportanus* Pfeiff.	HLF	M/C/J	Zo	Im
62	Epiphyllum phyllanthus **(L.) Haw.** *	HLC	M/C/J	Zo	8474
63	Lepismium cruciforme **(Vell.) Miq.** *	HLC	M/C/J	Zo	8388
64	*Lepismium lumbricoides* (Lem.) Barthlott	HLC	M/C/J	Zo	8386
65	Lepismium warmingianum **(K. Schum.) Barthlott** *	HLC	C	Zo	8390
66	*Rhipsalis baccifera* (J.S. Muell.) Stearn	HLC	M/J	Zo	8387
67	*Rhipsalis campos-portoana* Loefgr.	HLC	M	Zo	8389
68	*Rhipsalis cereuscula* Haw.	HLC	C/J	Zo	8383
69	Rhipsalis pilocarpa **loefgr.**∞	HLC	C	Zo	8385
70	Rhipsalis floccosa **Salm-Dyck ex Pfeiff.** *	HLC	M/C	Zo	8380
71	Rhipsalis paradoxa **(Salm-Dyck ex Pfeiff.) Salm-Dyck** *	HLC	M	Zo	8382
72	*Rhipsalis teres* (Vell.) Steud.	HLC	M/C/J	Zo	8384
73	*Rhipsalis trigona* Pfeiff.	HLC	M/C/J	Zo	8381
COMMELINACEAE					
74	*Tradescantia albiflora* Kunth	HLA	M/C/J	Zo	8391
DRYOPTERIDACEAE					
75	*Elaphoglossum lingua* (C. Presl) Brack.	HLC	M	An	8393
76	*Elaphoglossum glabellum* J.Sm.	HLF	M	An	8480
77	*Elaphoglossum glaziovii* (Fée) Brade	HLF	M	An	8479
78	*Elaphoglossum ornatum* (Mett. ex Kuhn) Christ	HLF	M	An	8476
79	*Polybotrya cylindrica* Kaulf.	HMS	M	An	8477
80	*Stigmatopteris caudata* (Raddi) C. Chr.	HLA	M	An	8478
GESNERIACEAE					
81	Codonanthe devosiana **Lem.** *	HLC	M	Zo	Im
82	Codonanthe gracilis **(Mart.) Hanst.** *	HLC	M	Zo	8491
83	*Nematanthus striatus* (Handro) Chautems	HLC	M	Zo	8489
84	Sinningia douglasii **(Lindl.) Chautems** *	HLC	M	Zo	8490
MARCGRAVIACEAE					
85	Marcgravia polyantha **Delpino** *	HMS	M	Zo	8502
ORCHIDACEAE					
86	*Acianthera recurva* (Lindl.) Pridgeon & M.W. Chase	HLC	J	An	8402
87	*Acianthera nemorosa* (Barb. Rodr.) F. Barros	HLC	J	An	8403
88	*Acianthera saundersiana* (Rchb.f.) Pridgeon & M.W.Chase	HLC	J	An	8401
89	*Anathallis obovata* (Lindl.) Pridgeon & M.W. Chase	HLC	J	An	8452
90	*Anathallis sclerophylla* (Lindl.) Pridgeon & M.W. Chase	HLC	M	Dïsp	8515
91	Baptistonia lietzei **(Regel) Chiron & V.P.Castro** *	HLC	C/J	An	8407
92	*Brasilidium* sp.	HLC	M	An	8398

Keep going...

Family	Species	EC	Area	Disp	Reg.
ORCHIDACEAE					
93	*Brasiliorchis gracilis* (Lindl.) R.B. Singer *et al.*	HLC	M	An	8396
94	*Brasiliorchis chrysantha* (Barb. Rodr.) R.B.Singer *et al.*	HLC	C	An	8397
95	*Bulbophyllum epiphytum* Barb. Rodr.	HLC	J	An	8374
96	*Bulbophyllum napellii* Lindl.	HLC	M	An	8414
97	Bulbophyllum plumosum **(Barb.Rodr.) Cogn.** *	HLC	J	An	8409
98	Bulbophyllum chloroglossum **Rchb.f. & Warm.** *	HLC	J	An	8408
99	*Campylocentrum aromaticum* Barb.Rodr.	HLC	M	An	8419
100	*Campylocentrum* cf. *grisebachii* Cogn.	HLC	M	An	Im
101	Capanemia micromera **Barb. Rodr.** *	HLC	M	An	8423
102	Catasetum fimbriatum **(C.Morren) Lindl.** *	HLC	M	An	8424
103	Catasetum atratum **lindl.**∞	HLC	M	An	8451
104	*Catasetum* sp.	HLC	M	An	8496
105	*Cattleya* sp.	HLC	M	An	Im
106	Coppensia varicosa **(Lindl.) Campacci** *	HLC	M	An	8406
107	*Cyclopogon multiflorus* Schltr.	HLA	M/C	An	8404
108	*Dichaea pendula* (Aubl.) Cogn.	HLC	M	An	8495
109	*Dichaea trulla* Rchb. f.	HLC	M	An	8470
110	*Encyclia oncidioides* (Lindl.) Schltr.	HLC	C	An	8466
111	*Encyclia patens* Hook.	HLC	M	An	8493
112	Epidendrum ansiferum **Rchb. f.** *	HLC	M	An	8373
113	*Epidendrum rigidum* Jacq.	HLC	C/J	An	8453
114	*Gomesa recurva* R. Br.	HLC	M	An	8410
115	Gomesa glaziovii **Cogn.** *	HLC	M	An	8411
116	*Gomesa* sp.	HLC	M	An	8417
117	*Grobya galeata* Lindl.	HLC	M	An	8405
118	*Lophiaris pumila* (Lindl.) Braem	HLC	M/C	An	8468
119	Miltonia flavescens**(Lindl.) Lindl.** *	HLC	C	An	8412
120	*Miltonia* sp.	HLC	M	An	Im
121	*Notylia longispicata* Hoehne & Schltr.	HLC	M	An	8416
122	*Octomeria crassifolia* Lindl.	HLC	J	An	8472
123	*Octomeria grandiflora* Lindl.	HLC	M	An	8418
124	*Octomeria palmyrabellae* Barb. Rodr.	HLC	J	An	8442
125	*Octomeria gracilis* Lodd. ex Lindl.	HLC	J	An	8440
126	*Oeceoclades maculata* (Lindl.) Lindl.	HLA	C/J	An	8400
127	Ornithocephalus myrticola **Lindl.** *	HLC	J	An	8443
128	*Polystachya concreta* (Jacq.) Garay & H.R. Sweet	HLC	M	An	8496
129	*Polystachya estrellensis* Rchb. f.	HLC	M/C/J	An	8467
130	*Polystachya foliosa* (Lindl.) Rchb.f.	HLC	C/J	An	8471
131	*Prosthechea glumacea* (Lindl.) W.E. Higgins	HLC	M	An	8399

| 132 | *Prosthechea* cf. *bulbosa* (Veli.) W.E. Higgins | HLC | M | An | Im |

Family	Species	EC	Area	Disp	Reg.
ORCHIDACEAE					
133	*Rodriguezia decora* (Lem.) Rchb.f.	HLC	M/C/J	An	8441
134	*Rodriguezia* sp.	HLC	J	An	8413
135	***Saundersia mirabilis* Rchb.f.'>'**	HLC	M	An	8494
136	*Scaphyglottis modesta* (Rchb. f.) Schltr.	HLC	M	An	8378
137	***Sophronitis cernua* Lindl. ***	HLC	C	An	Im
138	*Stelis deregularis* Barb. Rodr.	HLC	M	An	8473
PIPERACEAE					
139	*Peperomia alata* Ruiz & Pav.	HLC	M	Zo	8420
140	*Peperomia castelosensis* Yunck.	HLC	M	Zo	8421
141	*Peperomia catharinae* Miq.	HLC	M	Zo	8422
142	*Peperomia trineuroides* Dahlst.	HLC	C	Zo	8450
143	*Peperomia glabella* (Sw.) A. Dietr.	HLF	M/C/J	Zo	8475
144	*Peperomia pereskiifolia* (Jacq.) Kunth	HLC	M/C	Zo	8448
145	*Peperomia rotundifolia* (L.) Kunth	HLC	J	Zo	8449
146	*Peperomia tetraphylla* (G. Forst.) Hook. & Arn.	HLC	C/J	Zo	8447
147	*Peperomia trineura* Miq.	HLC	M/J	Zo	8446
148	*Peperomia urocarpa* Fisch. & C.A. Mey.	HLC	M	Zo	8492
POLYPODIACEAE					
149	*Campyloneurum acrocarpon* Fée	HLC	M	An	8529
150	*Campyloneurum centrobrasilianum* Lellinger	HLC	C	An	8526
151	*Campyloneurum nitidum* (Kaulf.) C. Presl	HLC	M/C	An	8531
152	*Campyloneurum repens* (Aubl.) C. Presl	HLC	M/C	An	8533
153	*Campyloneurum* sp.	HLC	J	An	8517
154	*Ceradenia albidula* (Baker) L.E. Bishop	HLC	M	An	8376
155	*Microgramma tecta* (Kaulf.) Alston	HLC	M/C/J	An	8524
156	***Microgramma crispata* (Fée) R.M.Tryon & A.F.Tryon ***	HLC	M	An	8530
157	*Microgramma lycopodioides* (L.) Copel.	HLC	M	An	8437
158	*Microgramma percussa* (Cav.) de la Sota	HLC	M	An	8392
159	*Microgramma persicariifolia* (Schrad.) C. Presl	HLC	M/C/J	An	8520
160	*Microgramma squamulosa* (Kaulf.) de la Sota	HLC	M/C/J	An	8534
161	*Microgramma vacciniifolia* (Langsd. & Fisch.) Copel.	HLC	M/C	An	8522
162	*Niphidium crassifolium* (L.) Lellinger	HLC	M	An	8516
163	*Pecluma filicula* (Kaulf.) M.G. Price	HLC	M/C/J	An	8527
164	*Pecluma* sp.	HLC	M	An	8375
165	***Pecluma truncorum* (Lindm.) M.G. Price ***	HLC	M	An	8535
166	*Pleopeltis astrolepis* (Liebm.) E. Fourn.	HLC	C/J	An	8519
167	*Pleopeltis hirsutissima* (Raddi) de la Sota	HLC	M/C/J	An	8521
168	*Pleopeltis macrocarpa* (Bory ex Willd.) Kaulf.	HLC	M	An	8435

169	*Pleopeltis pleopeltifolia* (Raddi) Alston	HLC	M/C/J	An	8525
170	*Pleopeltis squalida* (Veil.) de la Sota	HLC	M/C/J	An	8436

Table 53 - Continuation...

Family	Species	EC	Area	Disp	Reg.
POLYPODIACEAE					
171	*Serpocaulon catharinae* (Langsd. & Fisch.) A.R. Sm.	HLC	M	An	8528
172	*Serpocaulon fraxinifolium* (Jacq.) A.R. Sm.	HLC	M	An	8518
173	*Serpocaulon latipes* (Langsd. & Fisch.) A.R. Sm	HLC	M/C/J	An	8423
174	*Serpocaulon sehnemii* (Pic.-Serm.) Labiak & J.Prado	HLC	M	An	8532
PTERIDACEAE					
175	*Polytaenium cajenense* (Desv.) Benedict	HLC	M/C	An	8439
176	*Vittaria lineata* (L.) Sm.	HLC	M/C/J	An	8454

The richness of epiphytic species found in the Sorocaba/Médio Tietê hydrographic basin, when compared to studies with a similar methodology to this one, was lower than that observed in the results of Breier (2005) who, investigating four different Conservation Units distributed from east to west in the state of Sao Paulo, sampled 277 species. However, the results can be considered close to those observed by Kersten (2006), who studied the Alto Iguaçu Hydrographic Basin in Paranà and found 209 species in a field survey (349 spp. when including data from herbaria and other publications); by Bonnet et al. (2011), who surveyed three vegetation units of the Tibagi River in Paranà (a study similar to this one) and found 188 species of vascular epiphytes; and the results of Menini-Neto et al. (2009), who studied three Conservation Units in different forest formations and sampled 181 species of vascular epiphytes.

However, it can be considered lower than that observed in wetter forest formations, such as the Dense Ombrophilous Forest, for which some authors report a higher number of epiphyte species, e.g. Blum (2010) - 277 species; Kersten (2006) - 349 species, Fontoura et al. (1997) - 293 species, but similar to other studies carried out in the same forest, such as Petean (2009) - 159 species, Breier (2005) - 161 species, Petean (2003) - 97 species, Schütz-Gatti (2000) - 175 species and Hertel (1950) - 101 species.

When compared to the richness observed in surveys in Semideciduous Seasonal Forest, it is higher than the studies carried out by Rogalski and Zanin (2003) who found 70 species, by Giongo and Waechter (2004) who sampled 57 species, by Cervi and Borgo (2007) who found 56 species, by Dislich and Mantovani (1998), with 34 species, by Borgo et al. (2002), with 32 species, by Dettke et al. (2008), with 29 species, by Breier (2005), with 25 species and by Aguiar et al. (1981), who sampled 17 species.

When compared to floristic surveys of epiphytes carried out in Cerrado areas, the epiphytic community of the Sorocaba/Médio Tietê basin can be considered rich in species, especially when looking at the data from Breier (2005), who sampled 16 species, Ishara et al. (2008), seven species, Joanitti et al. (2010), 16 species and Bataghin et al. (2012b), who found 28 species.

The results reinforce the idea that the epiphytic community is dependent on atmospheric humidity (GENTRY; DODSON, 1987a), since the acquisition and storage of water are the most important factors for the establishment and survival of epiphytes (ZOTS; HIETZ 2001). In addition, the forest mosaic studied, which involves different phytophysiognomies and, within these, different environments, with variations in richness within each site, also contributed to the vascular epiphytic diversity recorded in this study, especially the record of forests with greater epiphytic diversity in the Upstream Area of the hydrographic basin (Figure 57).

In general, the vascular epiphyte survey carried out in the Sorocaba/Médio Tietê river basin recorded a considerable number of species, reaching sampling sufficiency (Figure 58). The epiphytic families with the highest species richness were: Orchidaceae (53 species), Bromeliaceae (42 species), Polypodiaceae (26 species), Cactaceae (13 species) and Araceae and Piperaceae (10 species each). Aspleniaceae, Dryopteridaceae, Gesneriaceae and Pteridaceae had six, six, four and two species respectively. The Begoniaceae, Blechnaceae, Commelinaceae and Marcgraviaceae families had only one species each. The presence of these families repeats the pattern found in other surveys of vascular epiphytes carried out in Brazil (BORGO; SILVA, 2003; ROGALSKI; ZANIN, 2003; GONÇALVES; WAECHTER, 2003; KERSTEN, 2006; DETTKE et al., 2008; GERALDINO et al., 2010; BONNET et al., 2011; LIMA et al., 2011).

The 14 families recorded in the Sorocaba/Médio Tietê watershed correspond to around 30% of the epiphytic families found in the Neotropics (GENTRY; DODSON, 1987a). Of the eight families common to the three areas of the river basin, seven (Araceae, Bromeliaceae, Cactaceae, Orchidaceae, Piperaceae, Polypodiaceae and Pteridaceae) are among the families with the highest proportion of epiphytic species in the world (MADISON, 1977; BENZING, 1990) and all eight, including the Commelinaceae family, are among the most common in the Brazilian Atlantic Forest (KERSTEN, 2010).

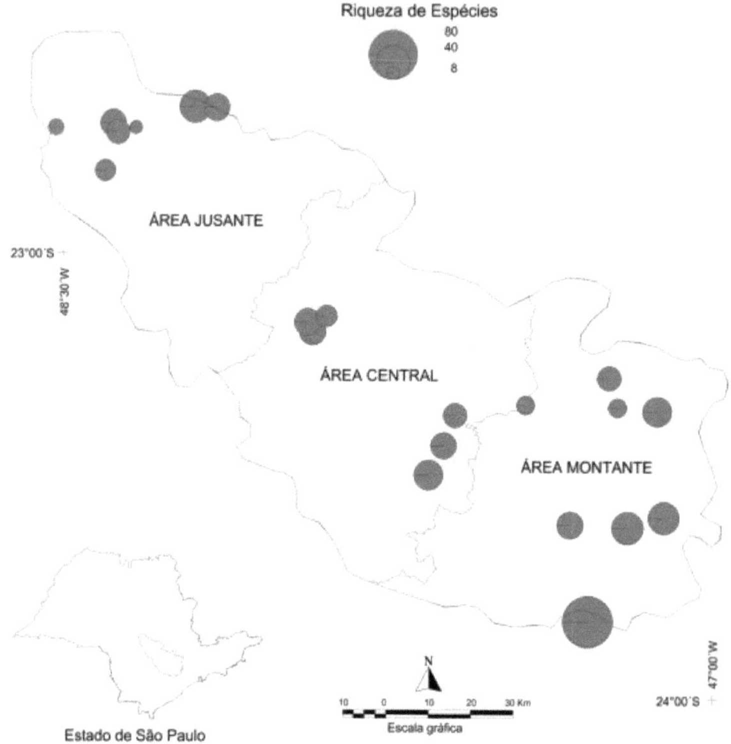

Figure 57: Distribution of vascular epiphyte richness in the sample sites in the three areas of the Sorocaba/Médio Tietê river basin.

The Orchidaceae, Bromeliaceae and Polypodiaceae families accounted for 121 species (68%) found in the floristic survey, a percentage very similar to that found by Kersten (2006). Furthermore, these families are considered to be the richest in epiphytes worldwide (MADISON, 1977; KRESS, 1986; GENTRY; DODSON, 1987b; BENZING, 1990). The Cactaceae family is also noteworthy in the basin area, because although it accounts for around 0.5% of the world's epiphytic species (MADISON, 1977; BENZING, 1990) and 3% of Brazilian epiphytes (KERSTEN, 2010), in the study area it had 13 species (7.3%). The resistance of the Cactaceae to periods of water stress and the great diversity of species of this family in the Neotropical region, where more than 1,400 species occur (HUNT et al., 2006), are responsible for the representativeness of this family in the area.

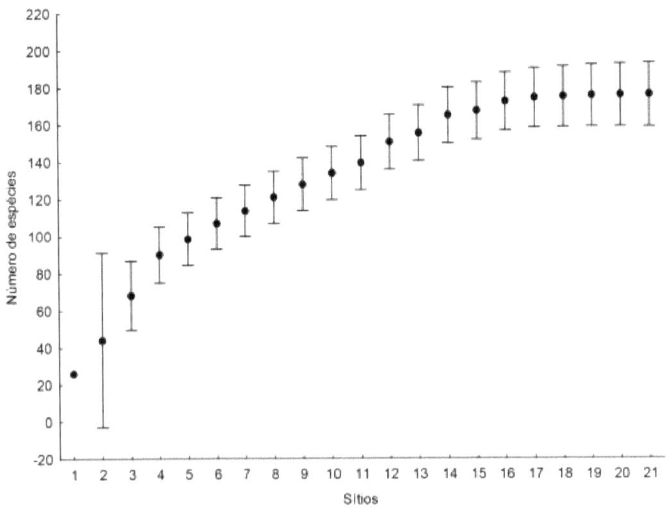

Figure 58: Estimated richness and confidence interval of vascular epiphytic species in the Sorocaba/Médio Tietê river basin.

The distribution of epiphytic species in the ecological categories (Figure 59), according to the relationship with the forophyte proposed by Benzing (1990), showed a predominance of characteristic holoepiphytes with 150 species (86%), followed by facultative holoepiphytes with 10 species (6%), secondary hemiepiphytes with six species (3%), accidental holoepiphytes, also with six species (3%) and primary hemiepiphytes with four species (2%). The predominance of characteristic holoepiphytes has been observed as a rule in studies carried out in Dense Ombrophilous Forest (BLUM, 2010; PETEAN, 2009; KERSTEN, 2006; BREIER, 2005; PETEAN, 2003; SCHÜTZ-GATTI, 2000; FONTOURA et al., 1997), in Semideciduous Seasonal Forest (PINTO et al., 1995; DISLICH; MANTOVANI, 1998; ROGALSKI; ZANIN, 2003; CERVI; BORGO, 2007; DETTKE et al., 2008; BATAGHIN et al., 2010), in Cerrado areas (BREIER, 2005; BATAGHIN et al., 2012b) and in other forest formations, such as Mixed Ombrophilous Forest (DITTRICH et al., 1999) and restinga forest areas (WAECHTER, 1992; KERSTEN; SILVA, 2001).

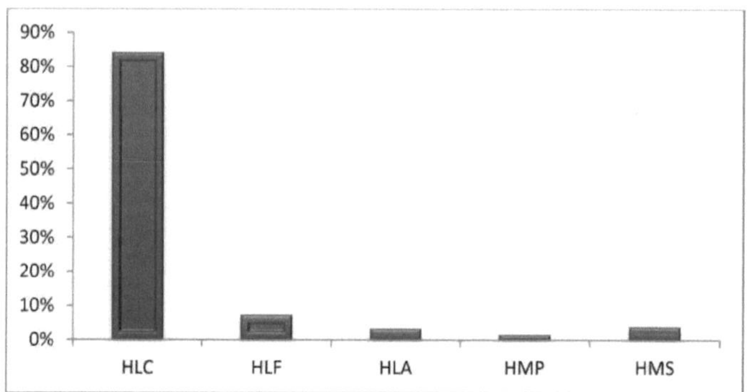

Figura 59: Distribution of the vascular epiphytic species of the Sorocaba/Médio Tietê Hydrographic Basin in the ecological categories proposed by Benzing (1990) - HLC: characteristic holoepiphytes; HLF: facultative holoepiphytes; HLA: accidental holoepiphytes; HMP: primary hemiepiphytes; HMS: secondary hemiepiphytes.

Dispersal strategy is an important factor in the success of epiphytic synopsis (GENTRY; DODSON, 1987a), and anemochory has predominated as a dispersal syndrome among epiphytic species (BENZING, 1987; BREIER, 2005; DETTKE et al., 2008; MENINI-NETO et al., 2009; GERALDINO et al., 2010). Of the 176 species sampled in this study, 55 showed zoochoric dispersal, while more than two thirds (121 spp.) are anemochoric. The results observed for the Sorocaba/Médio Tietê basin corroborate the hypothesis that two thirds of epiphytes have anemochoric dispersal (BENZING, 1987). In the basin 69% of the species showed anemochoric dispersal, while only 31% of the epiphytic flora showed zoochory as a dispersal syndrome. This high percentage of anemochory is a reflection of the large number of orchids, ferns and bromeliads (in the latter case, especially the genera *Tillandsia* and *Vriesea)* recorded in the basin.

The variation in the number of species between the different areas of the watershed, which was greatest in the Upstream Area (139 spp.), decreasing sharply and gradually in the Central (64 spp.) and Downstream Areas (56 spp.), was expected and can be explained by a combination of two basic factors that are important to the epiphytic community.), was expected and can be explained by the combination of two basic and important factors for the epiphytic community: i) the existing forest structure, since in the Upstream Area there are Ombrophilous Forests which are more shaded and contribute to the development of the vascular epiphytic community, in the Central Area there are predominantly Semideciduous Seasonal Forests and in the Downstream Area, a mixture of Semideciduous Seasonal Forest and Cerrado (in the latter two areas, the less dense vegetation allows more light in, which

178

reduces the humidity available to the vascular epiphytes, not to mention the very deciduousness of part of these forests, which exposes the epiphytic community to less favorable conditions for development); ii) the regularity of water availability (seasonality) in each area of the river basin, which is greater in the Upstream Area and less in the Central and Downstream Areas. For the Downstream Area, the irregular distribution of the wet season directly affects the vascular epiphytic community, because although it has average annual rainfall similar to the other areas of the basin, there is a marked dry period (in winter) during which the average monthly rainfall is below 60 mm, and can reach zero in some months (CEPAGRI, 2012). The variation in the number of epiphytic species according to forest type and water availability has been reported in several studies (GENTRY; DODSON, 1987b; BENZING, 1990; BARTHLOTT et al., 2001; NIEDER et al., 2001; ZOTZ et al., 2001; ARÉVALO; BETANCUR, 2006; BATAGHIN et al., 2010).

In the Sorocaba/Médio Tietê river basin, the epiphytic families common to all three Areas studied were: Araceae, Bromeliaceae, Cactaceae, Commelinaceae, Orchidaceae, Piperaceae, Polypodiaceae and Pteridaceae. Exclusive occurrences at the family level were only recorded in the Montante Area (Begoniaceae, Blechnaceae, Gesneriaceae, Marcgraviaceae).

The highest species richness, as previously mentioned, was found in the Orchidaceae, Bromeliaceae, Polypodiaceae, Cactaceae, Araceae and Piperaceae families, respectively. These families are responsible for 86.5% of the species recorded, although they occupy different positions within each area (Figure 60).

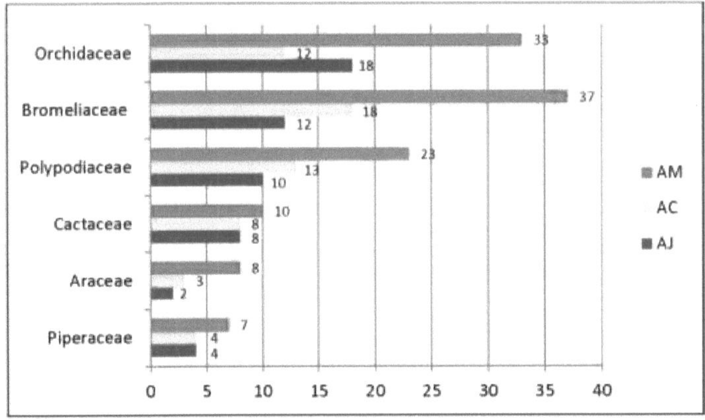

Figura 60: Species richness of six of the most representative families in the Sorocaba/Médio Tietê river basin: Upstream Area (AM), Central Area (AC) and Downstream Area (AJ). Sao Paulo, Brazil.

While Orchidaceae was the family with the highest richness in the Downstream Area (18 species), followed by Bromeliaceae and Polypodiaceae (12 and 10 spp., respectively), it appeared in second place in the Upstream Area (33 spp.) after Bromeliaceae (37 spp.) and only in third place in the Central Area (12 spp.) after Bromeliaceae (18 spp.) and Polypodiaceae (13 spp.). This reduction in the number of orchids in the Central Area is influenced by the numerous anthropogenic influences on the forest fragments in this area, a strong indication that this part of the hydrographic basin is the most in need of care in terms of environmental conservation. Cactaceae, Araceae and Piperaceae showed similar distribution patterns, being richer in the Upstream Area, and having lower richness in the Central and Downstream Areas.

The recording of a greater number of species belonging to the Orchidaceae family, followed by the Bromeliaceae, as recorded in this research, has been reported in several studies of epiphytes carried out in the Neotropics (INGRAM et al., 1996; NIEDER et al., 2000; BARTHLOTT et al., 2001; KERSTEN; SILVA, 2002; BORGO; SILVA, 2003; ROGALSKI; ZANIN, 2003; GIONGO; WAECHTER, 2004; KROMER et al., 2005; ALVES et al.m 2008; KERSTEN; KUNIYOSHI, 2009; MENINI-NETO et al., 2009; BLUM et al., 2011). However, in two of the three areas of the hydrographic basin studied, the Bromeliaceae were predominant, an unexpected but not uncommon result, especially in drier areas or those that have suffered greater human interference. Menini-Neto et al. (2009), studying different areas in Minas Gerais, observed in one of the areas that the richness of the Bromeliaceae was greater than that of the Orchidaceae, and attributed this variation to the environmental conditions present in that seasonal forest. In addition, Dettke et al. 2008, studying epiphytes in an impacted area, identified a small number of Orchidaceae species, while Bromeliaceae and Polypodiaceae had more species. The greater richness of Polypodiaceae in relation to Orchidaceae was also recorded in the Central Area of this river basin. The dominance of other families was also observed in the Neotropics by Arévalo and Betancur (2004) and Benavides et al. (2006), both of whom identified the Araceae family in areas of Colombian Guyana and Colombian Amazonia, respectively.

It is important to note that most of the studies in which Orchidaceae is highlighted as the richest family were carried out in moist forests or Ombrophilous Forests, while the results of Menini-Neto et al. (2009) and Dettke et al. (2008) were obtained in Seasonal Forests. This same behavior seems to be repeated in the three areas of the hydrographic basin studied, where in the Upland Area (Ombrophilous Forest) the Orchidaceae family is more numerous and in the other two areas (mixed Seasonal Forest and Cerrado) this family was less representative.

In terms of conservation, the Bromeliaceae family had the highest number of species threatened with extinction, considering the three large areas of the river basin, with 22 epiphytic species (Figure 61). This was followed by the Orchidaceae family with 13 species under some degree of threat and the Cactaceae family with 6 species.

The fact that 50 species were found under different degrees of threat in the forest fragments of this watershed (Table 53) highlights the importance of this area and the role it plays in conserving vascular epiphytic diversity for the Cerrado and Atlantic Rainforest. In addition, the accelerated rate of destruction of these plant formations and the high biological richness of both the Cerrado and the Atlantic Rainforest place them as priorities for biodiversity conservation at a global level (MYERS et al., 2000).

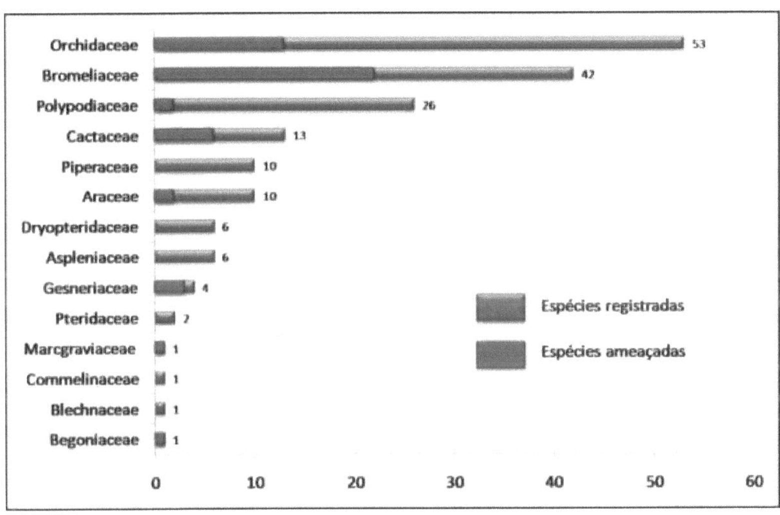

Figura 61: Distribution of epiphytic vascular species under some degree of threat of extinction in different families in the Sorocaba/Médio Tietê basin.

Of the 66 genera found, 21 were common to all three areas, while 28 were exclusive to the Upstream Area, two to the Downstream Area and one to the Central Area. Of the 176 species recorded, more than 56% (100 spp.) belong to the 16 most diverse genera, i.e. less than 1/4 of the genera recorded in the watershed. The most diverse genera were Vriesea (14 spp.) and Tillandsia (13 spp.), followed by Peperomia (10 spp.), Rhipsalis (8 spp.), Microgramma (7 spp.), Philodendron, Asplenium and Aechmea (6 spp.), Campyloneurum and Pleopeltis (5 spp.), Anthurium, Billbergia, Elaphoglossum, Bulbophyllum, Octomeria and Serpocaulon (4 spp.). All five of the most diverse genera were recorded in the three areas of the watershed.

181

In all three areas of the river basin, the genus *Tillandsia* was the richest. In the Upstream Area it had 12 species, in the Central Area six species and in the Downstream Area five species. This genus appears among the richest genera in several surveys of vascular epiphytes in Brazil (KERSTEN; SILVA, 2002; BORGO; SILVA, 2003; GONÇALVES; WAECHTER, 2003; ROGALSKI; ZANIN, 2003; GIONGO; WAECHTER, 2004; LINSINGEN et al. 2006, MENINI-NETO et al., 2009; GERALDINO et al., 2010, BONNET et al. 2011). This can be explained by the wide distribution of the genus *Tillandsia* in the Neotropics (VERSIEUX; WENDT, 2007; APG III, 2009), by its anemochoric dispersal syndrome, which facilitates its spread over large areas (MARTINELLI et al., 2008) and also due to the presence of mechanisms or adaptations that favor epiphytic habit, resisting variations in luminosity, temperature and humidity (BENZING, 1990; HIETZ; HIETZ-SEIFERT, 1995). However, various studies carried out in moist forests have shown that the genus Tillandsia is less expressive than other genera (KERSTEN, 2006; PETEAN, 2009; BLUM et al., 2011; LIMA et al. 2011).

Another highlight in the Montante Area was the genus *Vriesea,* which appeared as the second richest, with 11 species. This genus is usually more tolerant of shaded areas, although it has been recorded in areas with higher light intensity (BONNET; QUEIROZ 2006). In the Central and Downstream Areas, species from other genera with greater resistance to periods of water deficit also stand out, such as *Rhipsalis* and *Peperomia*, which, being succulent, resist the seasonality characteristic of drier forests such as the Seasonal Forests and the Cerrado, as well as having, in the case of *Peperomia, a* pantropical distribution (APG III, 2009).

All five of the richest genera in the watershed (Vriesea, *Tillandsia, Peperomia, Rhipsalis* and *Microgramma*) are widely distributed in Brazilian forests (FORZZA et al. 2013), but this does not contradict the importance of the floristic diversity found in each of the watershed areas.

The analysis of floristic similarity between the three areas of the hydrographic basin, calculated using the Jaccard coefficient (Downstream/Central Area = 0.4523, Downstream/Mountain Area = 0.2037 and Central/Mountain Area = 0.25) is illustrated in the Venn diagram (Figure 62), and shows areas with distinct floras, despite the proximity or contiguity between the areas.

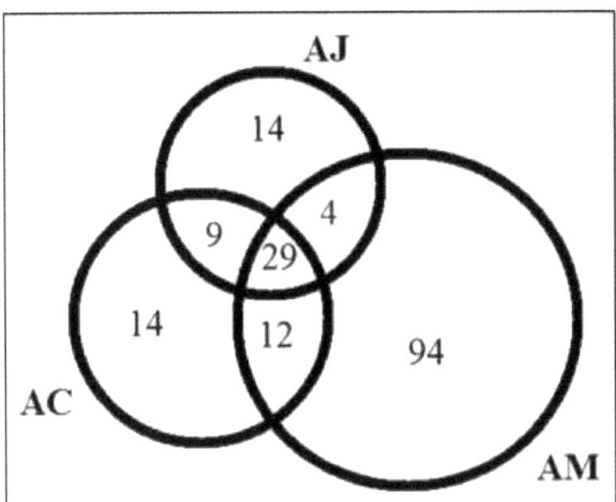

Figure 62: Venn diagram showing the common and shared species between the three areas of the Sorocaba/Médio Tietê river basin, Sao Paulo, Brazil.

There is a relatively small overlap of epiphytic species between the areas, with 29 species common to all three areas (Table 53). In addition to the common species in the basin, the Downstream and Central Areas showed the greatest floristic similarity, with an overlap of nine more species: Aechmea apocalyptica, Tillandsia funckiana, Rhipsalis cereuscula, Baptistonia lietzei, Epidendrum rigidum, Oeceoclades maculata, Polystachya foliosa, Peperomia tetraphylla and Pleopeltis astrolepis. However, between the areas with the least similarity (downstream and upstream), there are only four additional overlapping species: Philodendron appendiculatum, Vriesea bituminosa, Rhipsalis baccifera and Peperomia trineura.

The great similarity between the Central Area and the Downstream Area (38 species and 23 genera in common) was to be expected, as these are nearby areas that share physiognomic characteristics (Semideciduous Seasonal Forest). In addition, the climatic and microclimatic conditions in these two areas are quite similar, especially due to the irregular distribution of rainfall during the year (they have a well-defined dry period), which may also be one of the main reasons for the lower diversity in these two areas compared to the Montante Area. Gentry and Dodson (1987) emphasize the low tolerance of epiphytic species to periods of low water availability. In addition, for Kersten (2006) aridity excludes the competitiveness of most vascular epiphytic species.

The results also suggest, as expected, that the proximity between the areas is an important factor and tends to add to the existing phytophysiognomic similarities, as there is a greater sharing of exclusive species between the areas that are close to each other (Upstream/Central

Area (12 spp.) and Central/ Downstream Area (9 spp.)) and a lower number of species being shared by the more distant areas within the hydrographic basin (Upstream/ Downstream Area - 4 spp.).

The variations in the similarity coefficient between the areas of the Sorocaba/Médio Tietê hydrographic basin, highest between the Central and Downstream Areas and lowest, respectively, between the Central and Upstream Areas and between the Upstream and Downstream Areas, can be explained by the proximity between the areas and by the existence of "enclaves" of Semideciduous Seasonal Forest (predominant in the Central Area) in its two adjacent areas, as well as the similarities in water seasonality in the areas of the hydrographic basin (more marked in the Jusante and Central Areas). Notably, this influence on vascular epiphytic diversity is greater in the Downstream Area. The low coefficient of similarity between the Upstream and

Downstream also corroborates this hypothesis. This can be seen as an indication of the importance of conserving these unique forest fragments for the epiphytic vascular community, when trying to maintain the diversity of this synonymy. Menini-Neto et al. (2009) also observed a low coefficient of similarity in similar and relatively close phytophysiognomic areas.

Statistical analysis revealed no significant difference in the abundance of vascular epiphytes between the Downstream, Central and Upstream Areas of the Sorocaba/Médio Tietê river basin. The values obtained were as follows: the analysis between Area

Downstream and the Central Area showed t = 0.112 and p = 0.456; between the Downstream Area and the Central Area, t = 0.112 and p = 0.456.

Upstream, t = -0.090 and p = 0.464; and between the Central Area and the Upstream Area t = 0.036 and p = 0.486. The absence of a significant difference between the abundances of the epiphytic communities that occur in the different areas of the basin indicates that this community is similar in terms of the number of individuals.

However, the statistical analysis applied to the presence/absence of species indicated that the Upstream Area is significantly different from the other two areas of the basin, but showed no significant difference between the Downstream and Central Areas of the Sorocaba/Médio Tietê hydrographic basin. The results obtained were: between the

Downstream and the Central Area t = 1.116 p = 0.133; between the Downstream Area and the Upstream Area, t = 9.976 and p = 0.0001; and between the Central Area and the Upstream Area t, = 8.580 and p = 0.0001.

This result reinforces the idea that the existing forest structure has an influence on the vascular epiphytic community, given that there is a physiognomic similarity between the forests that occur in the Central and Downstream Areas. However, the main factor responsible for this difference in species records is the regularity of the wet periods (water availability / seasonality) in each area of the watershed. The Upstream Area has more regular water availability, which can be seen in the area's characteristic climate, while the Central and Downstream Areas tend to have at least one period of water deficit (dry winter), which reduces the composition of species that can survive these variations.

The quantitative assessment of the vascular epiphytes of all the sites (core and its replicas) in the Central Area, the Downstream Area and the Upstream Area recorded 150 species, and is shown in Table 54. Although the presence of an important number of species from the Orchidaceae family is recorded, it is possible to notice a greater abundance of species that are resistant to periods of water deficit, such as some genera from the Polypodiaceae and Bromeliaceae families, especially *Microgramma, Pleopeltis* and *Tillandsia,* which account for around 50% of the epiphytic importance value (VIE). This is a reflection of the characteristic climate of this region, where in most of the river basin there is a predominance of well-defined seasons, especially with regard to the distribution of rainfall which is marked by limiting periods, especially the winter which tends to be drier, favoring the success of the species of these genera.

Table 54 - Vascular epiphytes of the Sorocaba/Médio Tietê Hydrographic Basin, classified according to the value of epiphytic importance - nr: absolute number of occurrences in the strata; far: absolute frequency in the strata; ni: absolute number of occurrences in the forophytic individuals; fai: absolute frequency in the forophytic individuals; vt (total value): sum of the abundance estimates; vie: value of epiphytic importance; grade: average grade obtained.

Species	nr	far	ni	fai	vt	vie	note
Pleopeltis pleopeltifolia	1168	13,36	399	36,94	1893	12,531	1,62
Microgramma squamulosa	915	10,47	280	25,93	1564	10,354	1,71
Microgramma vacciniifolia	565	6,46	171	15,83	1098	7,269	1,94
Tillandsia recurvata	647	7,40	222	20,56	1041	6,891	1,61
Tillandsia tricholepis	498	5,70	162	15,00	810	5,362	1,63
Pleopeltis hirsutissima	353	4,04	153	14,17	556	3,681	1,58
Pleopeltis squalida	248	2,84	81	7,50	483	3,197	1,95
Rhipsalis teres	232	2,65	101	9,35	468	3,098	2,02
Microgramma tecta	248	2,84	81	7,50	424	2,807	1,71

Species	nr	far	ni	fai	vt	vie	note
Lepismium cruciforme	184	2,10	75	6,94	364	2,410	1,98
Peperomia rotundifolia	158	1,81	79	7,31	345	2,284	2,18

Table 54 - Continuation...

Species	nr	far	ni	fai	vt	vie	note
Vriesea incurvata	207	2,37	67	6,20	342	2,264	1,65
Epiphyllum phyllanthus	164	1,88	82	7,59	305	2,019	1,86
Lepismium lumbricoides	147	1,68	67	6,20	285	1,887	1,94
Billbergia distachya	155	1,77	85	7,87	273	1,807	1,76
Microgramma persicariifolia	144	1,65	53	4,91	267	1,768	1,85
Aechmea bromeliifolia	113	1,29	62	5,74	235	1,556	2,08
Serpocaulon latipes	116	1,33	55	5,09	190	1,258	1,64
Tillandsia stricta	123	1,41	53	4,91	182	1,205	1,48
Rhipsalis cereuscula	83	0,95	39	3,61	170	1,125	2,05
Campyloneurum acrocarpon	117	1,34	56	5,19	169	1,119	1,44
Dichaea trulla	103	1,18	48	4,44	163	1,079	1,58
Gomesa recurva	83	0,95	36	3,33	147	0,973	1,77
Lophiaris pumila	80	0,92	48	4,44	146	0,967	1,83
Tillandsia geminiflora	95	1,09	36	3,33	130	0,861	1,37
Philodendron bipinnatifidum	62	0,71	31	2,87	128	0,847	2,06
Anthurium sellowianum	60	0,69	23	2,13	113	0,748	1,88
Campyloneurum nitidum	67	0,77	37	3,43	112	0,741	1,67
Philodendron propinquum	67	0,77	23	2,13	109	0,722	1,63
Polystachya foliosa	39	0,45	18	1,67	89	0,589	2,28
Serpocaulon catharinae	53	0,61	25	2,31	83	0,549	1,57
Vriesea altodaserrae	43	0,49	20	1,85	81	0,536	1,88
Elaphoglossum ornatum	56	0,64	30	2,78	75	0,496	1,34
Pecluma truncorum	43	0,49	17	1,57	69	0,457	1,60
Tillandsia funckiana	46	0,53	22	2,04	69	0,457	1,50
Vriesea carinata	47	0,54	19	1,76	68	0,450	1,45
Ornithocephalus myrticola	36	0,41	16	1,48	66	0,437	1,83
Vittaria lineata	41	0,47	19	1,76	61	0,404	1,49
Sinningia douglasii	35	0,40	13	1,20	58	0,384	1,66
Begonia fruticosa	35	0,40	12	1,11	53	0,351	1,51
Vriesea bituminosa	26	0,30	14	1,30	50	0,331	1,92
Peperomia urocarpa	28	0,32	19	1,76	49	0,324	1,75
Tradescantia albiflora	30	0,34	15	1,39	49	0,324	1,63
Vriesea fenestralis	28	0,32	10	0,93	48	0,318	1,71
Campyloneurum repens	28	0,32	14	1,30	46	0,305	1,64

Aechmea nudicaulis	20	0,23	13	1,20	45	0,298	2,25
Marcgravia polyantha	25	0,29	9	0,83	42	0,278	1,68
Codonanthe gracilis	20	0,23	5	0,46	42	0,278	2,10
Aechmea distichantha	26	0,30	14	1,30	41	0,271	1,58
Microgramma percussa	28	0,32	11	1,02	41	0,271	1,46
Vriesea gigantea	19	0,22	13	1,20	41	0,271	2,16
Varicose coppensia	22	0,25	11	1,02	39	0,258	1,77

Keep going...

Table 54 - Continuation...

Species	**nr**	**far**	**ni**	**fai**	**vt**	**vie**	**note**
Microgramma lycopodioides	23	0,26	8	0,74	39	0,258	1,70
Peperomia trineura	21	0,24	12	1,11	39	0,258	1,86
Polytaenium cajenense	28	0,32	12	1,11	38	0,252	1,36
Pecluma sp.	22	0,25	10	0,93	37	0,245	1,68
Neoregelia laevis	16	0,18	9	0,83	36	0,238	2,25
Serpocaulon fraxinifolium	22	0,25	9	0,83	36	0,238	1,64
Lepismium warmingianum	19	0,22	8	0,74	34	0,225	1,79
Peperomia glabella	18	0,21	11	1,02	34	0,225	1,89
Vriesea rodigasiana	22	0,25	11	1,02	33	0,218	1,50
Elaphoglossum glabellum	17	0,19	8	0,74	30	0,199	1,76
Baptistonia lietzei	16	0,18	9	0,83	29	0,192	1,81
Billbergia zebrina	17	0,19	11	1,02	29	0,192	1,71
Acanthostachys strobilacea	14	0,16	8	0,74	28	0,185	2,00
Asplenium scandicinum	18	0,21	9	0,83	28	0,185	1,56
Philodendron appendiculatum	15	0,17	8	0,74	28	0,185	1,87
Campylocentrum aromaticum	21	0,24	12	1,11	27	0,179	1,29
Octomeria crassifolia	2	0,02	2	0,19	26	0,172	13,00
Peperomia tetraphylla	13	0,15	5	0,46	26	0,172	2,00
Rodriguezia decorates	14	0,16	5	0,46	25	0,165	1,79
Notylia longispicata	21	0,24	12	1,11	24	0,159	1,14
Saundersia mirabilis	17	0,19	9	0,83	24	0,159	1,41
Niphidium crassifolium	13	0,15	7	0,65	22	0,146	1,69
Canistrum lindenii	10	0,11	8	0,74	21	0,139	2,10
Asplenium mucronatum	12	0,14	4	0,37	19	0,126	1,58
Billbergia porteana	11	0,13	8	0,74	19	0,126	1,73
Epidendrum rigidum	12	0,14	7	0,65	18	0,119	1,50
Cereus alacriportanus	8	0,09	4	0,37	17	0,113	2,13
Miltonia sp.	11	0,13	7	0,65	17	0,113	1,55

Anthurium comtum	9	0,10	4	0,37	16	0,106	1,78
Campylocentrum cf. *grisebachii*	13	0,15	7	0,65	16	0,106	1,23
Serpocaulon sehnemii	11	0,13	7	0,65	16	0,106	1,45
Anthurium acutum	11	0,13	6	0,56	15	0,099	1,36
Gomesa glaziovii	10	0,11	5	0,46	15	0,099	1,50
Tillandsia sp.	11	0,13	6	0,56	15	0,099	1,36
Tillandsia usneoides	9	0,10	4	0,37	15	0,099	1,67
Vriesea hieroglyphica	7	0,08	4	0,37	15	0,099	2,14
Asplenium sp.	8	0,09	3	0,28	14	0,093	1,75
Pleopeltis macrocarpa	9	0,10	4	0,37	14	0,093	1,56
Bulbophyllum cf. *plumosum*	7	0,08	4	0,37	13	0,086	1,86
Scaphyglottis modesta	7	0,08	4	0,37	13	0,086	1,86
Blechnum binervatum	8	0,09	4	0,37	12	0,079	1,50

Keep going...

Species	nr	far	ni	fai	vt	vie	note
Brasilidium sp.	7	0,08	5	0,46	12	0,079	1,71
Bulbophyllum chloroglossum	4	0,05	2	0,19	12	0,079	3,00
Acianthera saundersiana	4	0,05	2	0,19	11	0,073	2,75
Elaphoglossum lingua	6	0,07	5	0,46	11	0,073	1,83
Tillandsia araujei Mez	8	0,09	4	0,37	11	0,073	1,38
Epidendrum ansiferum	5	0,06	3	0,28	10	0,066	2,00
Octomeria gracilis	4	0,05	3	0,28	10	0,066	2,50
Peperomia castelosensis	6	0,07	3	0,28	10	0,066	1,67
Philodendron vargealtense	5	0,06	2	0,19	10	0,066	2,00
Tillandsia dura	7	0,08	4	0,37	10	0,066	1,43
Anthurium longifolium	5	0,06	4	0,37	9	0,060	1,80
Dichaea pendula	7	0,08	4	0,37	9	0,060	1,29
Peperomia catharinae	4	0,05	2	0,19	9	0,060	2,25
Rodriguezia sp.	6	0,07	3	0,28	9	0,060	1,50
Peperomia pereskiifolia	3	0,03	2	0,19	8	0,053	2,67
Bulbophyllum epiphytum	3	0,03	3	0,28	7	0,046	2,33
Oeceoclades maculata	12	0,14	8	0,74	7	0,046	0,58
Peperomia alata	4	0,05	2	0,19	7	0,046	1,75
Pleopeltis astrolepis	4	0,05	3	0,28	7	0,046	1,75
Vriesea platynema	4	0,05	3	0,28	7	0,046	1,75
Asplenium auritum	4	0,05	2	0,19	6	0,040	1,50
Campyloneurum centrobrasilianum	3	0,03	3	0,28	6	0,040	2,00
Grobya galeata	3	0,03	2	0,19	6	0,040	2,00

Species	nr	far	ni	fai	vt	vie	note
Asplenium pteropus	5	0,06	4	0,37	5	0,033	1,00
Cyclopogon multiflorus	4	0,05	2	0,19	5	0,033	1,25
Nidularium rutilans	2	0,02	1	0,09	5	0,033	2,50
Prosthechea glumacea	3	0,03	2	0,19	5	0,033	1,67
Rhipsalis baccifera	4	0,05	2	0,19	5	0,033	1,25
Rhipsalis trigona	3	0,03	1	0,09	5	0,033	1,67
Campyloneurum sp.	2	0,02	2	0,19	4	0,026	2,00
Catasetum fimbriatum	2	0,02	2	0,19	4	0,026	2,00
Catasetum atratum	2	0,02	2	0,19	4	0,026	2,00
Gomesa sp.	3	0,03	2	0,19	4	0,026	1,33
Octomeria palmyrabellae	2	0,02	1	0,09	4	0,026	2,00
Octomeria grandiflora	4	0,05	2	0,19	4	0,026	1,00
Pecluma filicula	2	0,02	1	0,09	4	0,026	2,00
Peperomia trineuroides	2	0,02	1	0,09	4	0,026	2,00
Philodendron corcovadense	3	0,03	1	0,09	4	0,026	1,33
Acianthera nemorosa	2	0,02	1	0,09	3	0,020	1,50
Asplenium pulchellum	2	0,02	1	0,09	3	0,020	1,50
Capanemia micromera	3	0,03	2	0,19	3	0,020	1,00
Species	**nr**	**far**	**ni**	**fai**	**vt**	**vie**	**note**
Elaphoglossum glaziovii	2	0,02	1	0,09	3	0,020	1,50
Encyclia patens	2	0,02	1	0,09	3	0,020	1,50
Polybotrya cylindrica	3	0,03	2	0,19	3	0,020	1,00
Prosthechea cf. *bulbosa*	2	0,02	1	0,09	3	0,020	1,50
Rhipsalis pilocarpa	3	0,03	3	0,28	3	0,020	1,00
Stelis deregularis	2	0,02	1	0,09	3	0,020	1,50
Tillandsia tenuifolia	2	0,02	1	0,09	3	0,020	1,50
Aechmea apocalyptica	1	0,01	1	0,09	2	0,013	2,00
Ceradenia albidula	2	0,02	1	0,09	2	0,013	1,00
Encyclia oncidioides	1	0,01	1	0,09	2	0,013	2,00
Nematanthus striatus	1	0,01	1	0,09	2	0,013	2,00
Philodendron eximium	1	0,01	1	0,09	2	0,013	2,00
Polystachya concreta	1	0,01	1	0,09	2	0,013	2,00
Rhipsalis campos-portoana	1	0,01	1	0,09	2	0,013	2,00
Stigmatopteris caudata	1	0,01	1	0,09	2	0,013	2,00
Polystachya estrellensis	1	0,01	1	0,09	1	0,007	1,00

The species with the highest importance value when considering all *the* quantitative sites in the Sorocaba/Médio Tietê Basin was *Pleopeltis pleopeltifolia* (Polypodiaceae) with an epiphytic importance value (EVI) of 12.531 and an average score of 1.62; this species occurred in 36.94% of the forophytes and 13.36% of the strata sampled. The second species with the highest importance value was *Microgramma squamulosa* (Polypodiaceae) with a VIE of 10.354 and an average score of 1.71, which was recorded in 25.93% of the forophytes and 10.47% of the strata. *Microgramma vacciniifolia* (Polypodiaceae) with a VIE of 7.269 and an average score of 1.94, occurring in 15.83% of the forophytes and 6.46% of the strata, was the third most important species. *Tillandsia recurvata* (Bromeliaceae) had a VIE of 6.891 and an average score of 1.61, occurring in 20.56% of the forophytes and 7.40% of the strata. *Tillandsia tricholepis* (Bromeliaceae) had a VIE of 5.362 and an average score of 1.63, occurring in 15% of the forophytes and 5.70% of the strata. These five species were responsible for more than 42% of the epiphytic importance value in the hydrographic basin studied.

The Polypodiaceae, Bromeliaceae, Cactaceae and Orchidaceae families, which are among the most frequently observed in Brazilian studies of Dense Ombrophilous Forests (BLUM, 2010; PETEAN, 2009; KERSTEN, 2006; BREIER, 2005; PETEAN, 2003; SCHÜTZ-GATTI, 2000; FONTOURA et al, 1997), in Semideciduous Seasonal Forest (DISLICH; MANTOVANI, 1998; ROGALSKI; ZANIN, 2003; GIONGO; WAECHTER, 2004; BREIER, 2005; DETTKE et al., 2008; BATAGHIN et al., 2010) and in Cerrado areas (BREIER, 2005; BATAGHIN et al., 2012b), were responsible for almost 90% of the epiphytic importance value. The Polypodiaceae family was the most important in the basin, with a VIE of 47.54, followed by Bromeliaceae with a VIE of 24.53, Cactaceae with a VIE of 10.97 and species from the Orchidaceae family, which accounted for 6.82% of the VIE in the Sorocaba/Médio Tietê basin.

Resistance to water deficit and/or temperature variation, both of which are present as a result of both the natural characteristics of the phytophysiognomies studied and anthropogenic changes in the forest remnants studied, may be responsible for the success of more cosmopolitan families, such as the Polypodiaceae. The Orchidaceae family is noteworthy because, although it had the highest number of species, it had a low importance value in this river basin. Factors such as greater sensitivity to climatic and microclimatic changes may be related to this fact; another possibility is that the species of this family suffer from the extraction of individuals from the forests, given their great interest and commercial value as ornamental plants.

The distribution of epiphytes in the strata, looking at all the sites in this watershed, highlighted the base of the canopy as the stratum with the highest epiphytic abundance (Figure 63), with an abundance value (AV) of 4954, followed by the inner canopy with an AV of 3393, the high stem with an AV of 3146 and the middle stem, low stem and outer canopy, with abundance values of 1790, 923 and 900, respectively.

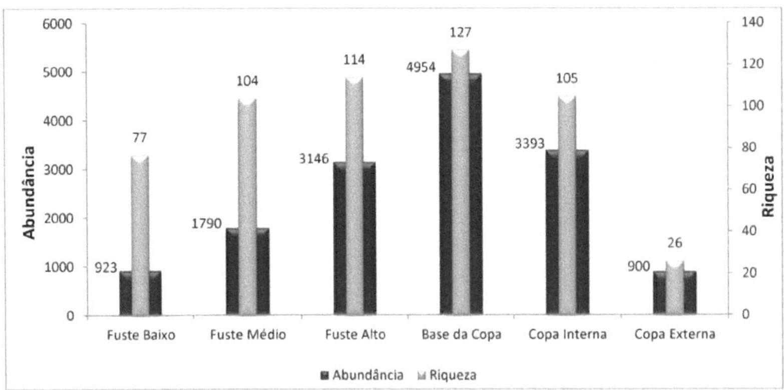

Figura 63: Distribution of the abundance and richness of epiphytic vascular species among forophytic strata in the Sorocaba/Médio Tietê Hydrographic Basin.

The vertical distribution of the epiphytic community in the Sorocaba/Médio Tietê basin, based on species abundance (Table 55), showed the following results: The low stem and outer canopy were significantly different from all the other strata, except for each other. The middle canopy differed significantly from all the other strata. This pattern of distribution of the abundance of vascular epiphytes in the hydrographic basin is typical of areas under anthropogenic influence or in intermediate successional stages, such as that recorded at the Rèplica II and Rèplica III sites in the Montante Area. Although the greater abundance of the middle stem (Figure 63) suggests a better conservation condition (given that the lower strata are more abundant in these forests), the lack of a significant difference between the lower stem and the outer canopy, as well as a very similar number of individuals, a behavior observed in impacted forest fragments, are indicative of the fragile general state of conservation of these remnants in this watershed. In addition, the great abundance of lower strata in the Core Site of the Upland Area may be "camouflaging" the real state of conservation of the watershed, which in most of the sites shows alterations to its original condition.

The concentration of abundance in the intermediate strata was expected and was observed in all (quantitative) sites in the watershed. This can be attributed to the greater availability of area in this part of the host trees, as well as the favorable environment for the installation and development of epiphytic individuals, either by the presence of branches that favor the anchoring of propagules (especially those of anemochoric dispersal), or by the retention of suspended soil that contributes to the retention of moisture fundamental to this community.

Table 55: Analysis of the distribution of the richness and abundance of vascular epiphytes among the forophytic strata in the Sorocaba/Médio Tietê river basin.

^^^^Richness Abundance^-^^	Beam Bass	Beam Medium	High stem	World base	Cup Internal Cup	Cup External
Low stem		**0,001471**	**9,28E-06**	**6,52E-10**	**9,56E-04**	**6,01E-10**
Medium stem	**0,008**		0,1962	**0,001637**	0,9009	**1,17E-19**
High stem	**4,44E-04**	**0,025**		0,05954	0,243	**2,81E-24**
Cup base	**0,0001**	**0,002**	0,068		**0,002463**	**2,45E-31**
Internal Cup	**0,004**	**0,044**	0,411	0,127		**4,31E-20**
External Cup	0,48	**0,035**	**0,001**	**1,50E-04**	**0,006**	

Epiphytes are irregularly distributed throughout the forophytes, with the number of individuals and species varying vertically, as well as differences in floristic composition (STEEGE; CORNELISSEN, 1989; BROWN, 1990; WAECHTER, 1992). This was proven, in part, by the significant differences between the outer canopy and all the other strata (Table 55), although it should be noted that only two of the 26 species recorded in this region (Figure 63) do not occur in all the forophyte strata, i.e. the species present in the outer canopy generally show great environmental plasticity within forest fragments.

In the case of the lower canopy, even though it has a similar number of individuals to the outer canopy, it has almost three times as many species. In this stratum, in addition to the "generalist" species, a very rich epiphytic community develops which is also very sensitive to environmental changes, since both the number of species and the number of individuals are reduced with environmental changes (and consequent microclimatic changes). This same observation also applies to the mid-stem, so that the species that inhabit the lower part of the forest are the most sensitive and vulnerable to microclimatic changes, whether or not they come from anthropogenic alterations to the environment.

The vertical evolution of vascular epiphytes occurred (and still occurs) through the exchange of lower spaces in search of more light, even under limited conditions for acquiring water and nutrients (KIRA; YODA, 1989; BENZING, 1990). Evidence of this is the large number of species recorded in the intermediate strata, regardless of the type of forest or climate studied (Figure 64), especially at the base of the canopy, which was significantly different from all strata except the high stem (Table 55). The base of the canopy was able to support the largest number of individuals among the strata, evidence of the favorable environment for vascular epiphytism that it represents (FREIBERG, 1996; NIEDER et al., 1999; ACEBEY; KROMER, 2001).

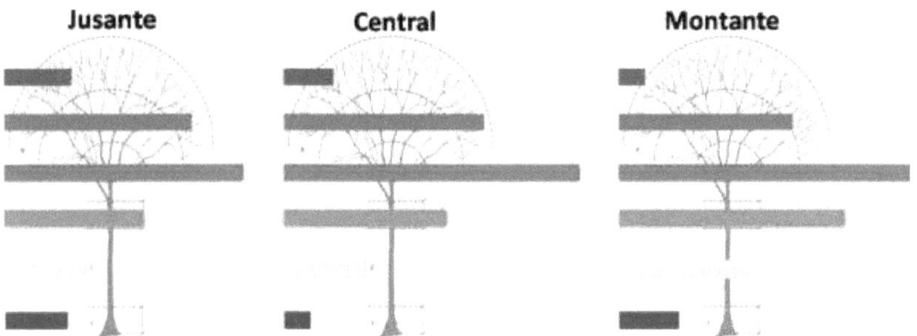

Figura 64: Vertical distribution of vascular epiphyte abundances in three large areas of the Sorocaba/Médio Tietè river basin.

The intermediate region shows greater floristic similarity to each other than to the extremes of the forophyte (Figure 65). The greater similarity between the base of the canopy and the inner canopy, and between these and the high stem, is related to the presence of places that tend to accumulate suspended soil (branch insertion points), retaining nutrients and moisture, which favors the establishment and development of epiphytes (BENZING, 1990; KERSTEN, 2006). Another factor that may be at work is the search for a balance between water acquisition and competition for light within the forest.

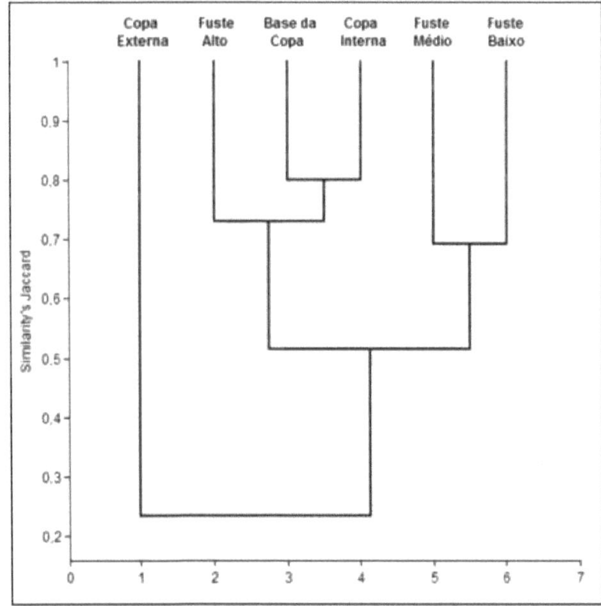

Figura 65: Dendrogram (UPGMA) of Jaccard's similarity between strata in the Sorocaba/Médio Tietê river basin.

The similarities in the distribution of the abundance of the epiphytic vascular community between the low and mid stems can be attributed to the occurrence of species that have little resistance to periods of water deficit, such as the Araceae, or even species that prefer the shadier areas of the forest, such as the Piperaceae. The distribution of abundance and the low similarity of the outer canopy in relation to the other strata may be related to the light intensity, which is the opposite of what occurs in the low and medium canopies; in addition, extreme temperature events may affect this stratum more than the other strata studied. The epiphytic community that occurs in the outer canopy tends to be more resistant to water deficit, benefiting from abundant light, both of which should have a strong effect on epiphytic species in the Sorocaba/Médio Tiete basin.

The statistical analysis applied to the vertical distribution of vascular epiphyte abundances in the different areas of the hydrographic basin revealed that this distribution occurs differently between the Downstream and Upstream Areas, and between the Central and Upstream Areas, with $p = 0.0001$ for both comparisons. The analysis between the vertical distribution of the Central and Downstream Areas showed no significant difference ($p > 0.05$). The results obtained in the individual statistical analysis for the different zones (strata) are presented below (p-values ≤ 0.05 indicate a significant difference).

Low stem -Between the Downstream Area and the Central Area, $p = 0.314$; between the Downstream Area and the Upstream Area, $p = 0.007$; and between the Central Area and the Upstream Area, $p = 0.023$.

Medium stem - Between the Downstream Area and the Central Area, $p = 0.200$; between the Downstream Area and the Upstream Area, $p = 0.0001$; and between the Central Area and the Upstream Area, $p = 7.87^{-04}$.

High stem - Between the Downstream Area and the Central Area, $p = 0.321$; between the Downstream Area and the Upstream Area, $p = 0.0001$; and between the Central Area and the Upstream Area, $p = 0.0001$.

Cup Base - Between the Downstream Area and the Central Area, $p = 0.489$; between the Downstream Area and the Upstream Area, $p = 0.0001$; and between the Central Area and the Upstream Area, $p = 0.0001$.

Internal Cup - Between the Downstream Area and the Central Area, $p = 0.453$; between the Downstream Area and the Upstream Area, $p = 0.0001$; and between the Central Area and the Upstream Area, $p = 0.0001$.

External Cup - Between the Downstream Area and the Central Area, $p = 0.397$; between the

Downstream Area and the Upstream Area, p = 0.080; and between the Central Area and the Upstream Area, p = 0.131.

There is a notable difference in the vertical distribution of vascular epiphytes between the Upstream Area and the other areas of the watershed, while there is no difference in the vertical distribution of the vascular epiphytic community between the Downstream Area and the Central Area. In the analysis applied individually to each of the strata, the outer canopy of the Upstream Area was the only stratum that showed no significant difference in relation to the same zone in the other areas. This observation indicates that the

Montante, as well as harboring a different community of vascular epiphytes, has a different occupation or distribution on the forophytes. This result further highlights the similarities, already pointed out above, between the Downstream Area and the Central Area of the Sorocaba/Médio Tietê hydrographic basin. Factors that may undoubtedly have contributed are the climatic and vegetational similarities between these two areas. For the Montante Area, the difference in the vertical distribution of vascular epiphytes compared to the other areas was expected, as it is an odd area with phytophysiognomic and climatic characteristics that contribute to vascular epiphytism.

CHAPTER 5

FINAL CONSIDERATIONS

Biodiversity conservation is one of today's biggest challenges, given the high level of human disturbance to natural ecosystems. One of the main consequences of these disturbances is the fragmentation of ecosystems, especially forests. Most forest remnants, especially in intensively cultivated landscapes, are found in the form of small, highly disturbed, isolated, little-known and poorly protected fragments (VIANA, 1995).

The study of the epiphytic vascular community is extremely important, since epiphytes are excellent indicators of the state of conservation of forests, and can reveal the disturbances exerted by humans in these forests (SOTA, 1971; WOLF, 2005; BARTHLOTT et al., 2001; DETTKE et al., 2008; KERSTEN; KUNIYOSHI, 2009; BONNET; QUEIROZ, 2006; BATAGHIN et al., 2010; KERSTEN, 2010). In addition, epiphytes function as biological indicators of the successional stage of forests (MEIRA, 1997; WOLF, 2005; BARTHLOTT et al., 2001; BATAGHIN et al., 2008).

Considering the analyses carried out by this study, which verified the richness of epiphytic species found in the Sorocaba/Médio Tietê river basin and the structure of their distribution and abundance among the sites studied, there is a multiplicity of agents responsible for or modifying this synopsis in the three areas of the river basin, especially in the Semideciduous Seasonal Forest/Cerrado ecotone of the Downstream Area and the Semideciduous Seasonal Forest of the Central Area of the basin, which makes the vascular epiphytic community vulnerable. The preservation of these areas must necessarily take all these factors into account, especially with regard to the relationship between man and natural ecosystems. With this in mind and based on the objectives of this research, specific considerations were drawn up for each of the areas in the watershed.

For the Downstream Area, the fact that it has the lowest number of vascular epiphytic species among the three areas of the hydrographic basin and the occurrence of epiphytic species common to anthropic environments (BATAGHIN et al., 2008; DETTKE et al., 2008; BATAGHIN et al., 2012a), especially the dominance of species from genera such as Tillandsia and Pleopeltis (Bromeliaceae and Polypodiaceae, respectively), as well as their distribution in the fragments and on the forbs, are indicative of the existence of environmental degradation. Engwald et al. (2000), studying the diversity, spatial distribution and dynamics of processes in epiphytic vascular vegetation in different types of forest, reported the loss of 55% of species in altered areas and/or areas with secondary vegetation. In addition, the climatic variations in this area of the basin, especially the occurrence of a period of marked water deficit, has an influence on epiphytic diversity. The absence or low structural complexity, especially in the case of Sitios Rèplica I and Quantitativo III, means that the epiphytic community, as well as being not very diverse, is dominated by a few species, most of which are not very abundant.

Although in this Downstream Area around 2/3 of the vascular epiphytes showed an anemochoric dispersal syndrome, as expected (BENZING, 1987), there is a strong tendency for this proportion to decrease in small forest fragments (< 30 ha), or when these forest fragments show low structural complexity, especially due to anthropogenic pressures. In some sites in this area with these characteristics, the ratio of anemochoric to zoochoric epiphytic species was close to 50%.

For the ecotone between the Semideciduous Seasonal Forest and the Cerrado, there was no difference in the abundance of epiphytes between the Core Site (Conservation Unit) and Replicate Sites I, II and III, which are forest remnants in the midst of an agricultural matrix and which suffer various types of anthropogenic pressure. However, there were significant differences in the presence/absence of vascular epiphytic species between the Core Site and five of the six replicas analyzed, adding to the statistical results 31 vascular epiphytic species that only occur in forest fragments that are not protected as PAs. It is therefore possible to state that the epiphytic community occurring in the downstream area of the Sorocaba/Médio Tietê river basin is heterogeneous in terms of its specific composition, and that most of the epiphytic species are at risk of disappearing locally. The results also suggest that the possible microclimatic variations at the Core and Replica Sites do not influence the abundance of epiphytes, but do influence the composition and distribution of vascular epiphyte species between the sites.

The record of a statistically significant difference only between the vertical distribution of Sitio Core and Sitio Rèplica I, in this area, added to the greater abundance in the intermediate forophytic strata, both for Sitio Core and the replica sites, indicates that, although distinct communities of vascular epiphytes are occurring in the different forest fragments, there is similarity in their behavior or distribution, in that they occupy the vertical strata of the forophytes analyzed in a similar way; in addition, the vascular epiphytic community that develops in the lower strata is fragile and tends to reduce in number of species and individuals when the forest remnants are altered.

One fact that deserves attention is the presence of a greater number of epiphytic species at Sitio Réplica III, a conserved forest remnant on private property. At this site, the greater success of the epiphytic community can be attributed to the "buffer" effect offered by the very well preserved tree vegetation in the area. Less anthropogenic interference favors the presence of a microclimate suitable for vascular epiphytism. An example of this is the presence of a considerable number of orchid species, which are more sensitive to environmental variations, as well as the greater number of species and individuals in the lower strata. This lower level of human interference has a direct impact on the preservation of forophytes, especially the larger ones, which not only facilitate the occurrence of vascular epiphyte species, due to the greater area available and the longer time for the species to settle, but also because they serve as a source of propagules for the younger trees.

Species from the Orchidaceae family are more sought after by collectors because of their ornamental potential. The fact that this family is richer and more abundant in the Downstream Area than in the Central Area may also be an indication of less anthropogenic interference in this part of the basin than in the Central Area. This may also indicate that, in the Downstream Area of the Sorocaba/Médio Tietê basin, the main effect regulating vascular epiphyte diversity is the climatic gradient, marked by colder, drier winters and wetter summers, and in second order would be anthropogenic disturbances to forest remnants, although these are ultimately responsible for reducing the diversity of vascular epiphytes.

This area plays a very important role in the landscape of the Sorocaba/Médio Tietê basin, as it brings together a set of biotic and abiotic factors that condition a characteristic vegetation, which represents a formation of great genetic and conservation value. The low similarity, as measured by Jaccard's coefficient, between the epiphytes observed in this area of the basin and those found in other studies carried out in similar vegetation formations in Brazil (in general $J < 0.5$, except in comparison with a study carried out in the area of this same river basin) and the presence of distinct epiphytic communities between the Core Site and the Replica Sites further highlights the importance and need to protect these forest remnants.

In the same way as was recorded in the Downstream Area of the basin, for the Central Area the occurrence of epiphytic species common to anthropogenic environments (BATAGHIN et al., 2008; DETTKE et al., 2008), especially the dominance of species from the Polypodiaceae and Bromeliaceae families, and the very similar floristic composition of vascular epiphytes between the sites analyzed, can be considered evidence of the environmental degradation of the area. Notably, altered areas and/or secondary vegetation can reduce epiphytic diversity by more than 50% (ENGWALD et al., 2000; BARTHLOTT et al., 2001). In addition, the low or even zero structural complexity of the forest fragments in the Central Area has become clear, which means that the epiphytic community, in addition to being little diverse in this part of the basin (or at least less diverse than would be expected), takes much longer to (re)establish, either because of the absence of medium and large-sized phorophytes, which are essential for the establishment and/or dispersal of epiphytes, or because of the unfavorable microclimatic conditions, caused mainly by anthropogenic action such as the opening of trails (roads) and the suppression of the understory for agricultural activities.

In the Central Area of this hydrographic basin, there was a considerable reduction in the expected proportion (2/3) of anemochoric/zoochoric epiphytic species, with around 55% of species recorded as anemochoric and 45% as zoochoric. This reduction in the number of anemochoric species can be attributed to the occurrence of forest fragments that are totally dispersed in the anthropogenic matrix and also to the low structural complexity of the fragments studied. The distance from other forest fragments may represent a barrier to anemochoric dispersal and consequent colonization of new areas by the species. The low structural complexity does not provide support for the development of anemochoric species that are more demanding in microclimatic terms, as is the case with orchids; in addition, this low structural complexity contributes to the development of species that are more adapted to conditions of great climatic variability, such as the Cactaceae, which have zoochoric dispersal. It should also be noted that this reduction in the expected proportion of anemochoric/zoochoric epiphytic species was also observed in the Downstream Area, in isolated forest fragments with low structural complexity.

One aspect that stands out in the Central Area of the Sorocaba/Médio Tietê river basin is the lack of difference, in terms of both floristic composition and abundance, between the Core Site, a Conservation Unit of over 5,000 ha, and the Replica Sites, isolated and altered forest remnants in the midst of an agricultural matrix that suffer from various types of anthropogenic pressure. These data show that the epiphytic community that occurs in the Central Area of the Sorocaba/Médio Tietê river basin is homogeneous, both in terms of specific composition and abundance. Contrary to what might be expected, the data suggest that the possible microclimatic variations existing in the Core Site and its Replicas do not influence the distribution of vascular epiphyte species between these sites, or that the Core Site in the Central Area has a high degree of anthropogenic influence (due to the history of the area) which does not allow for a differentiation in richness or abundance between this and the Replicas Sites, hypothetically more affected by human action.

However, the recording of 43 vascular epiphytic species exclusively in unprotected forest fragments suggests a concern about the conservation of vascular epiphytes in this area of the basin, and raises a reflection on the importance of actions and public policies towards the preservation and conservation of small forest fragments, especially those close to water bodies that have notable microclimatic characteristics that favor vascular epiphytes.

The absence of any significant difference between the vertical distribution of the Core Site and the Replica Sites in this area of the hydrographic basin, together with the greater abundance of epiphytes in the intermediate forophytic strata and the reduced number of species and individuals in the lower strata, both for the Core Site and the Replica Sites, indicates a vascular epiphytic community characteristic of impacted forest fragments subjected to strong microclimatic variations. In addition, the record of a greater number of individuals and species in the intermediate strata in this area may be associated not only with the presence of a greater number of branch insertion points, which favors the establishment and development of this community, but also with the accumulation of suspended soil which contributes to moisture retention, given that the lack of structural complexity in forest fragments extremely reduces water availability. The similarity in the behavior and distribution of epiphytic species at all the sites in the Central Area reinforces the need to adopt conservation measures in this part of the basin.

In the Central Area, Sitio Règlica III stands out, even though it has the smallest area (31 ha) of all the sites studied in this part of the basin, it had the highest alpha diversity of vascular epiphytes. Although it is common sense, and reported by several studies, that changes in the landscape negatively influence the diversity and abundance of vascular epiphytes, small forest remnants should not be ignored in conservation actions, as it seems that forest fragmentation may have isolated or even restricted numerous epiphytic populations in these forest fragments. This observation once again highlights the importance of conserving small areas of forest, even if they have been impacted, or even individual trees isolated in the landscape, so that they can shelter the vascular epiphytic community.

The low similarity, as measured by Jaccard's coefficient, between the vascular epiphyte communities observed in this Central Area of the watershed and the communities found in other studies carried out in similar vegetation formations in Brazil (in general J < 0.3, except in comparison with two other studies), in addition to the presence of a large number of epiphytic species that occur exclusively in the Replica Sites (43 spp. recorded in forest fragments that are not protected in the form of a Conservation Unit) further reinforces the importance of protecting these forest remnants.

Observing the richness of epiphytic species found in the Montante Area and the structure of their distribution among the sites studied, it can be said that this area is the best conserved within the Sorocaba/Médio Tietê hydrographic basin, although there are a number of agents modifying the forest landscape in the Semideciduous Seasonal Forest/Dense Ombrophilous Forest ecotone. This situation requires attention if the vascular epiphytic community is not to become vulnerable, as is the case in other areas of this basin. The preservation of forests in the Montante Area must consider all the factors that act on natural ecosystems, but pay special attention to those that concern the relationship between man and these ecosystems, since the loss of forest trees can mean the concomitant loss of a natural genetic heritage that is often overlooked, as is the case with vascular epiphytes.

Even though the richness of the Montante Area is greater than that of the other areas of the basin studied, we can see the occurrence of epiphytic species common to anthropogenic environments, especially the dominance of species from genera such as Pleopeltis and Microgramma (Polypodiaceae), and Tillandsia (Bromeliaceae), which is indicative of the existence of some degree of environmental degradation, especially in the unprotected forest fragments (Replica Sites). In addition, in the case of the Replica II and Replica III sites, the lack of structural complexity in the forest, due to the presence of anthropogenic interference, means that the epiphytic community, as well as being not very diverse, is dominated by a few species belonging to genera common to impacted environments, while the majority of species are not very abundant.

The Upstream Area of the Sorocaba/Médio Tietê basin was the only area that maintained the expected proportion of 2/3 (BENZING, 1987) among the vascular epiphytic species dispersed by anemochory/zoochory. Notably, this percentage of anemochory is a reflection of the large number of orchids, ferns and bromeliads (in the latter case, especially the genera *Tillandsia* and *Vriesea)* recorded in the area. However, it is worth noting the better vegetation conditions in this area, which means that there is greater connectivity between the forest fragments and also the presence of greater structural complexity in the forest fragments.

There was no significant difference between the abundance of epiphytes that occur in the Core Site (Conservation Unit) and those that occur in the Replica Sites I, II and III, despite the fact that the latter are smaller forest remnants which, in most cases, suffer from various types of anthropogenic pressure. However, there were significant differences in the presence/absence of species between the Core Site and its replicas ($p < 0.0001$), which allows us to state that the epiphytic community that occurs in the Upstream Area of the hydrographic basin is heterogeneous in terms of its specific composition, especially in the Conservation Unit studied. The results of this research suggest that there are climatic and microclimatic variations between Core and Replica Sites. Although these do not influence the abundance of epiphytes, they do influence the composition and distribution of vascular epiphyte species, which vary between sites.

One fact that deserves attention in this area of the basin is the presence of a greater number of epiphytic species in Sitio Core, a Conservation Unit. The greater success of the epiphytic community at this site can certainly be attributed, in addition to the peculiar climatic characteristics, to the well-preserved tree vegetation in the area. Although there is some anthropogenic interference, the presence of a microclimate favorable to vascular epiphytism means that a considerable number of orchid species, which are more sensitive to environmental variations, occur there. It should be noted that human interference is less intense inside protected areas than in unprotected fragments, which has a direct impact on the preservation of larger forophytes, which facilitate the occurrence of vascular epiphyte species, both because of the larger area available and the longer time for the species to settle, and because they serve as a source of propagules for the (re)settlement of adjacent areas.

Species from the Orchidaceae family are extracted from the wild by humans because of their ornamental interest and commercial value. Although this family is the richest in the Upstream Area, the number of species was lower than expected, even though it is richer than in the Downstream Area and the Central Area. This may be an indication of anthropogenic interference (predatory collecting) on this family in this part of the basin, because although the main regulators of vascular epiphyte diversity are climatic and microclimatic gradients, resulting especially from water availability, anthropogenic disturbances to forest remnants are responsible for reducing the biodiversity of vascular epiphytes.

This area of the watershed was the only one in which the vertical distribution of the vascular epiphyte community detected in the Core Site was significantly different from that recorded in Replica Sites I, II and III. This further highlights the importance of the Core Site (Conservation Unit) in this part of the basin, as in addition to the greater number of vascular epiphytic species it houses, it has an important singularity in the vertical occupation of these over the host trees.

Similarly to what was observed in the other areas of the basin, in both the Core Site and the Replica Sites, the intermediate forophyll strata showed greater abundance, benefiting from the presence of a greater number of branch insertion points, which favors the accumulation of suspended soil, and also from the microclimate favorable to vascular epiphytism. However, the lower strata showed a different pattern of vertical distribution of the individuals to that observed in the other Areas (Downstream and Central), especially due to the greater abundance of the low stem in relation to the outer canopy, which has proved to be a good indicator of the state of conservation of the forest, since even in areas of dry forest (Replica

Site III of the Downstream Area), when the forest remnant is well conserved, it showed this same form of vertical distribution.

Another important observation is that in the Upstream Area, the outer canopy showed the lowest abundance among the strata and also recorded fewer individuals than this same stratum in the Central and Downstream Areas. This may indicate that the epiphytic vascular community that occupies the outer canopy in this part of the watershed is more sensitive to environments with greater climatic stress, either due to water availability or factors such as humidity and/or temperature.

The low similarity, as measured by Jaccard's coefficient, between the vascular epiphyte community observed in the Montante Area of the basin and those found by other studies carried out in similar vegetation formations in Brazil (J < 0.31), as well as the presence of a large number of epiphytic species, reinforce the importance of protecting and conserving forest areas in this part of the Sorocaba/Médio Tietê river basin.

In general, the areas studied in the Sorocaba/Médio Tietê basin play a very important role in the regional landscape, as they bring together a set of biotic and abiotic factors that condition a characteristic vegetation, which represents a formation of great genetic and conservation value. In view of the results of this research, which verified the richness of epiphytic species found in forest fragments, the structure of their distribution among the sites studied, and also identified a multiplicity of modifying agents in the Sorocaba/Médio Tietê basin, it can be concluded that it is vulnerable, and its preservation must necessarily take all these factors into account, especially with regard to the relationship between man and natural ecosystems.

The presence of distinct epiphytic communities, both between the different areas of the basin (Downstream, Central and Upstream) and between the sites (Core and Replicas) highlights the importance and need to protect the remaining forest.

The behavior or vertical distribution of the vascular epiphytes differed only in the upstream area of the watershed, indicating that the communities that develop in the Downstream and Central Areas occupy the spaces on the forophytes in a similar way. In addition, the Upstream Area was the only one in which the epiphytic vascular community of the Core Site (UC) was distributed differently over the host trees, while in the Downstream and Central Areas the epiphytic community is established in a similar way over the forophytes, regardless of whether or not it is protected as a Conservation Unit. The vertical distribution of vascular epiphytes on forophytes can be used as a useful tool to characterize the state of conservation of forest fragments. In conserved areas, the lower strata (especially the low stem) tend to show greater diversity and abundance than in non-preserved forest fragments. In addition, the three sites that showed this form of vertical distribution, namely the Core Site in the Upland Area, and Sites Rèplica II in the Upland Area and Rèplica III in the Downstream Area showed the highest diversity indices (H^r). However, Site Rèplica III in the Central Area, which had a better preserved understory, had the highest alpha diversity in this part of the watershed.

The differences, in terms of species diversity, between the forest fragments, especially between the Core and Replica Sites in each area of the basin, as well as between the three areas of the river basin, in addition to the low coefficient of similarity observed between the total vascular epiphytic flora occurring in this river basin and similar surveys in southern and southeastern Brazil, reinforce the importance of preserving forest fragments, regardless of their area or the phytophysiognomy involved. In addition, the presence of 50 species listed under some degree of threat in the Red Book of Brazilian Flora (2013) provides a strong argument for the conservation of forest fragments in the Sorocaba/Mèdio Tietè basin.

BIBLIOGRAPHICAL REFERENCES

ACEBEY A. and KROMER T. 2001. Diversidad y distribución vertical de epifitas en los alrededores del campamento rio Eslabón y de la laguna Chalalàn, Parque Nacional Madidi, Dpto. La Paz, Bolivia. **Revista de la Sociedad Boliviana de Botànica** 3: 104123.

AGUIAR L.W., CITADINE-ZANETTE V., MARTAU L. and BACKES A. 1981. Floristic composition of vascular epiphytes in an area located in the municipality of Montenegro and Triunfo, Rio Grande do Sul, Brazil. **Iheringia** (Série Botânica) 28:55-93.

ALVES R.J.V. and KOLBEK J. 2009. Summit vascular flora of the Serra de Sao José, Minas Gerais, Brazil. **Check List** 5(1):35-73

ALVES R.J.V., KOLBEK J. and BECKER J. 2008. Vascular epiphyte vegetation in rocky savannas of southeastern Brazil. **Nord J Bot** 26:101-117

ANGIOSPERM PHYLOGENY GROUP-APG III. 2009. An update of the Angiosperm Phylogeny Group classification for the orders and families of flowering plants: APG III. **Botanical Journal of the Linnean Society.** 161:105-121. http://dx.doi.org/10.1111/j.1095-8339.2009.00996.x

ARÉVALO R. and BETANCUR J. 2004. Diversidad de epifitas vasculares en cuatro bosques del sector suoriental de la Serrania de Chiribiquete, Guayana Colombiana. **Caldasia** 26(2):359-380

ARÉVALO R. and BETANCUR J. 2006. Vertical distribution of vascular epiphytes in four forest types of the Serrania de Chiribiquete, Colombian Guayana. **Selbyana** 27:175-185

AYRES M., AYRES JÙNIOR M., AYRES D.L. and SANTOS A.S. 2007. BIOESTAT 5.9 - Statistical applications in the bio-medical sciences. Mamiraua NGO. Belém.

BARROS F., HAMEDE M.C.H., MELO M.M.R.F., LOPES E.A., JUNG- MENDAÇOLLI S.L., KIRIZAWA M., MUNIZ C.F.S., MAKINO-WATANABE H. CHIEA S.A.C. and MELHEM T.S. 2002. The phanerogamic flora of PEFI: composition, affinities and conservation. In. Bicudo D.C., Forti M.C., Bicudo C.E.M. (eds.). **Parque Estadual das Fontes do Ipiranga (PEFI): A conservation site that resists the urbanization of Sao Paulo.** Editora Secretaria do Meio Ambiente do Estado de Sao Paulo. Sao Paulo.

BARTHLOTT W., SCHMIT-NEUERBURG V., NIEDER J. and ENGWALD S. 2001. Diversity and abundance of vascular epiphytes: a comparison of secondary vegetation and primary montane rain forest in the Venezuelan Andes. **Plant Ecology** 152:145-156.

BATAGHIN F.A, MULLER A., PIRES J.S.R., BARROS F., FUSHITA A.T. and SCARIOT E.C. 2012b. Richness and vertical stratification of vascular epiphytes in the

Jatai Ecological Station: Cerrado Area in Southeastern Brazil. **Hoehnea** 39(4):615- 626

BATAGHIN F.A. 2009. **Distribution of the vascular epiphyte community at different sites in the Ipanema National Forest, Iperó, SP, Brazil.** Master's dissertation. Federal University of Sao Carlos, Sao Carlos. Sao Paulo.

BATAGHIN F.A., BARROS F. and PIRES J.S.R. 2010. Distribution of the vascular epiphyte community in sites under different degrees of disturbance in the Ipanema National Forest, Sao Paulo, Brazil. **Revista Brasileira de Botânica** 33(3):531-542.

BATAGHIN F.A., FIORI A. and TOPPA R.H. 2008. Edge effect on vascular epiphytes in Mixed Ombrophilous Forest, Rio Grande do Sul, Brazil. **O Mundo da Saùde** 32:329-338.

BATAGHIN F.A., PIRES J.S.R. and BARROS, F. 2012a. Vascular epiphytism in edge and interior sites in Semideciduous Seasonal Forest in Southeastern Brazil. **Hoehnea** 39(2):235-245

BENAVIDES A.M., Wolf J.H.D. and DUIVENVOORDEN J.F. 2006. Recovery and succession of epiphytes in upper Amazonian fallows. **Journal of Tropical Ecology** 22:705-717

BENNET B.C. 1986. Patchiness, diversity, and abundance relatioships of vascular Epiphytes. **Selbyana** 9:70-75

BENZING D.H. 1986. The vegetative basis of vascular epiphytism. **Selbyana** 9:23-43.

BENZING D.H. 1987. Vascular epiphytism: taxonomic participation and adaptive diversity. **Annals of the Missouri Botanical Garden** 74(2):183-204

BENZING D.H. 1990. **Vascular epiphytes.** Cambridge University Press, Cambridge.

BENZING D.H. 1995. The physical mosaic and plant variety in forest canopies. **Selbyana**16:159-168.

BERNARDI S. and BUDKE J.C. 2010. Structure of epiphytic synteny and edge effect in a transition area between Semideciduous Seasonal Forest and Mixed Ombrophilous Forest. **Floresta** 40:81-92.

BIERREGAARD R.O., LOVEJOY T.E., KAPOS V., SANTOS A.A. and HUTCHINGS W. 1992. The biological dynamics of tropical rain forest fragments. **BioScience** 42:859-866.

BLUM C.T. 2010. **The epiphytic vascular and herbaceous terricolous components of the**

Ombrophilous Dense Forest along an altitudinal gradient in the Serra da Prata, Paranà. PhD thesis in Forest Engineering - Nature Conservation. Federal University of Paranà.

BLUM C.T., RODERJAN C.V. and GALVAO F. 2011. Floristic composition and altitudinal distribution of vascular epiphytes of the Ombrophilous Dense Forest in Serra da Prata, Morretes, Paranâ, Brazil.**Biota Neotropica11**(4):141-159

BONNET A. and QUEIROZ M.H. 2000. Considerations on epiphytic bromeliads as indicators of degraded forests. Desterro Environmental Conservation Unit, Santa Catarina Island. In. **Proceedings of** the II Brazilian Congress of Conservation Units. Rede Nacional Pro Unidades de Conservaçao e Fundaçao o Boticârio de Proteçao à Natureza, Campo Grande, v.2, p.217-221.

BONNET A. and QUEIROZ M.H. 2006. Vertical stratification of epiphytic bromeliads in different successional stages of the Ombrophilous Dense Forest, Santa Catarina Island, Santa Catarina, Brazil. **Revista Brasileira de Botânica** 29:217-228.

BONNET A., CURCIO G.R., LAVORANTI, O.J. and GALVAO F. 2011. Epiphytic vascular flora in three vegetation units of the Tibagi River, Paranà, Brazil. **Rodriguésia** 62(3):491-498.

BONNET A., LAVORANTI O.J. and CURCIO, G.R. 2009. Relationships between vascular epiphytes and environmental factors in the Araucaria biodiversity corridor, Paranâ. **InAnais** do IX Congresso de Ecologia do Brasil - Ecologia e o futuro da biosfera. Sao Lourenço, v.1, p.1-4.

BORGO M. and SILVA S.M. 2003. Vascular epiphytes in fragments of Mixed Ombrophilous Forest, Curitiba, Paranâ, Brazil. **Revista Brasileira de Botànica** 26(3):391- 401

BORGO M., SILVA S.M. and PETEAN M.P. 2002. Vascular epiphytes in a remnant of semideciduous seasonal forest, municipality of Fènix, PR, Brazil. **Acta Biologica Leopoldensia** 24:121-130.

BRAUN-BLANQUET J. 1979. **Phytosociology: bases for the study of vegetation communities.** H. Blume Edic. Madrid.

BREIER T.B. 2005. **Vascular epiphytism in forests of southwestern Brazil.** Doctoral thesis. State University of Campinas. Campinas.

BROWN A.D. 1990. Epiphytism in the montane forests of El Rey National Park, Argentina: floristic composition and distribution pattern. **Revista de Biologia Tropical** 38:155-166.

BRUMMITT R.K. and POWELL C.E. 1992. **Authors of plant names.** 1st ed., Royal Botanic Gardens, Kew.

CALLAWAY R.M., BROOKER R.W., CHOLER P., KIKVIDZE Z., LORTIE C.J., MICHALET R., PAOLINI L., PUGNAIRE F.I., NEWINGHAM B., ASCHEHOUG E.T., ARMAS C., KIKODZE D.E. and COOK, B.J. 2002. Positive interactions among alpine plants increase with stress. **Nature** 417:844-848.

CALLAWAY R.M., REINHART K.O., TUCKER S.C. and PENNINGS, S.C. 2001. Effects of epiphytic lichens on host preference of the vascular epiphyte Tillandsia usneioides. **Oikos** 94:433-441.

CBH-SMT and FABH-SMT. 2008. **Fundamentals of charging for the use of water resources in the Sorocaba and Mèdio Tietê Basin.** Material prepared by the Technical Group for Charging for Water Use. Sorocaba, Sao Paulo.

CEPAGRI 2012. Center for Meteorological and Climatic Research Applied to Agriculture. **Koeppen's climate classification for the state of Sao Paulo**. Campinas. Sao Paulo.

CERVI A.C. and BORGO M. 2007. Vascular epiphytes in the Iguaçu National Park, Paranà (Brazil). Preliminary survey. **Fontqueria** 55(51): 415-422.

CLARK K.L., NADKARNI N.M., SCHANFER D. and GHOLZ H.L. 1998. Atmospheric deposition and net retention of ions by the capony in a tropical montane forest, Monteverde, Costa Rica. **Journal of Tropical Ecology** 14:27-45

CONSERVATION INTERNATIONAL DO BRASIL. Fundaçao SOS Mata Atlântica, Fundaçao Biodiversitas, Instituto de Pesquisas Ecológicas, Secretaria do Meio Ambiente do Estado de Sao Paulo, Instituto Estadual de Florestas-MG. 2000. **Assessment and priority actions for the conservation of the biodiversity of the Atlantic Rainforest and Southern Fields.** Secretariat of Biodiversity and Forests of the Ministry of the Environment. Brasilia

COTTAN G. and CURTIS J.T. 1956. The use of distance measures in phytosociological sampling. **Ecology** 37(4):451-460.

COUTINHO L.M. 1962. Contribution to the knowledge of the ecology of tropical rain forest. **Boletim da Faculdade de Ciências e Letras da Universidade de Sao Paulo, Botânica** 18:11-219.

DETTKE G.A., ORFRINI A.C. and MILANEZE-GUTIERRE M.A. 2008. Floristic composition and distribution of vascular epiphytes in an altered remnant of Semideciduous Seasonal Forest in Paranà, Brazil. **Rodriguésia** 59:859-872.

DISLICH R. and MANTOVANI W. 1998. Flora of vascular epiphytes of the University City Reserve "Armando de Salles Oliveira" (Sao Paulo, Brazil). **Boletim de Botânica da**

Universidade de Sao Paulo 17:61-83.

DITT E.H. 2002. **Forest fragments of the Paranapanema portal.** Annablume, Sao Paulo.

DITTRICH V.A.O., KOZERA C. and SILVA, S.M. 1999. Floristic survey of vascular epiphytes in Barigüi Park, Paranà, Brazil. **Iheringia** (Sèrie Botànica) 52:1122.

DURIGAN G., SIQUEIRA M.F., FRANCO G.A.D.C. and RATTER, J.A. 2006. Selection of priority fragments for the creation of cerrado conservation units in the state of Sao Paulo. **Revista do Instituto Florestal**, Sao Paulo. 18:23-37.

ELLENBERG H. and MUELLER-DUMBOISD. 1966. "Tentative physiognomic - ecological classification of plant formations of the earth". **Separate part of Ber. Geobot. Inst.** ETH, Zurich, 1965/1966.

EMBRAPA. 1999. **Brazilian Soil Classification System.** National Soil Research Center. Rio de Janeiro.

ENGWALD S., SCHIMIT-NEUERBURG V. and BARTHLOTT, W. 2000. Epiphytes in rain forest of Venezuela - diversity and dynamics of a biocenosis. In.**Results of worldwide ecological studies** (S.W. Breckle, B. Schweizer & U. Arndt, eds.). Proceedings of the first symposium by the AFW Foundation. Günter Heimbach, Hoheneim, p.425-434.

ePIC - **Electronic Plant Information Center** 2013. www.rbgkew.org.uk/epic (Last accessed on 15/08/2013)

FAVERO O.A., NUCCI J.C. and BIASI M. 2004. Potential natural vegetation and mapping of current vegetation and land use in the Ipanema National Forest, Iperó/SP: Conservation and environmental management. **RA'E GA - O Espaço Geogràfico em Anâlise** 8:55-68.

FIDALGO O. and BONONI V.L.R. 1989. **Techniques for collecting, preserving and herborizing botanical material.** Botanical Institute, Sao Paulo.

FLORES-PALACIOS A. and GRACIA-FRANCO G. 2006. The relationship between tree size and epiphytic species richness: testing four different hypotheses. **Journal of Biogeography** 33:323-330.

FONTOURA T. 2001. Bromeliaceae and other epiphytes - stratification and resources available to animals in the Jacarepià State Ecological Reserve, Rio de Janeiro. **Bromélia** 6:33-39.

FONTOURA T., SYLVESTRE L.S., VAZ A.M.S. and VIEIRA, C.M. 1997. Vascular epiphytes, hemiepiphytes and hemiparasites of the Macaé de Cima ecological reserve. In. H.C. Lima and R.R. Guedes-Bruni, (eds.). **Serra de Macaé de Cima: Floristic diversity and**

conservation of the Atlantic Rainforest. Editora Jardim Botànico, Rio de Janeiro, p.89-101.

FORZZA R.C. et al. 2013 **Lista de Espécies da Flora do Brasil 2013** in http://floradobrasil.jbrj.gov.br/

FREIBERG, M. 1996. Spatial distribution of vascular epiphytes on three emergent canopy trees in French Guiana. **Biotropica** 28: 345-355.

GENTRY A.H. 1982. Neotropical floristic diversity: phytogeographical connections between Central and South America, Pleistocene climatic fluctuations, or an accident of the Andean orogeny? **Annals of the Missouri Botanical Garden** 19:149-156.

GENTRY A.H. and DODSON C.H. 1987a. Diversity and biogeography of neotropical vascular epiphytes. **Annals of the Missouri Botanical Garden** 74:205-233.

GENTRY A.H. and DODSON, C.H. 1987b. Contribution of non trees species to the richness of a tropical rain forest. **Biotropica** 19:149-156.

GERALDINO H.C.L., CAXAMBU M.G. and SOUZA D.C. 2010. Floristic composition and community structure of vascular epiphytes in an ecotone area in Campo Mourao, PR, Brazil. **Acta Botânica Brasilica** 24(2):469-482

GIONGO C. and WAECHTER J.L. 2004. Floristic composition and community structure of vascular epiphytes in a gallery forest in the Central Depression of Rio Grande do Sul. **Revista Brasileira de Botânica** 27:563-572.

GONÇALVES C.N. and WAECHTER J.L. 2003. Floristic and ecological aspects of vascular epiphytes on isolated fig trees in the northern coastal plain of Rio Grande do Sul. **Acta Botânica Brasilica**17(1):89-100

GONÇALVES, C.N. and WAECHTER, J.L. 2002. Vascular epiphytes on isolated *Ficus organensis* specimens in the northern coastal plain of Rio Grande do Sul: Patterns of abundance and distribution. **Acta Botanica Brasilica** 16:429-441.

GRACIANO C., FERNANDEZ L.V. and CALDIZ, D.O. 2003. *Tillandsia recurvata* L. as a bioindicator of sulfur atmospheric pollution. **Ecologia Austral** 13:3-14.

GRIME J.P. 2001. **Plant strategies, vegetation processes, and ecosystem properties.**2nd ed, John Wiley, New York.

HADEL V.F. 1989. **The fauna associated with the bromeliad phytotelmata of the Juréia-Itatins Ecological Station (SP).** Master's dissertation. University of Sao Paulo. Sao Paulo-SP.

HAMMER 0., HARPER D.A.T., RYAN P.D. 2001. PAST: Paleontological Statistics software package for education and data analysis. **Palaeontologia Electronica** 4(1):1-9

HELBSING S., RIEDERER M. and ZOTZ, G. 2000. Cuticles of vascular epiphytes: Acient barriers for water loss after stomatal closure? **Annals of Botany** 86:765-769.

HENDERSON A. 1993. Literature on air pollution and lichens XXXVI. **Lichenologist** 25(2):191-202.

HERNADES-ROSAS J.I. 2001. Occupation of bearers by vascular epiphytes in a tropical rainforest of the Upper Orinoco, Edo. Amazonas, Venezuela. **Acta Cientifica Venezolana** 52:292-303.

HERTEL R.J.G. 1950. Contribution to the ecology of the epiphytic flora of the Serra do Mar (western slope) of Paranà. **Archives of the Museo Paranaense** 8:3-63.

HIETZ P. and HIETZ-SEIFERT U. 1995. Composition and ecology of vascular epiphyte communities along an altitudinal gradient in central Veracruz, Mexico. **Journal of Vegetation Science** 6: 487-498.

HUECK K. 1972. **The forests of South America: ecology, composition and economic importance.** Sao Paulo. Ed. Univ. Brasilia/Poligono.

HUNT D., TAYLOR N. and CHARLES G. 2006, The new cactus lexicon. DH Books, Milborne Port.

IBAMA. 2003. **Management Plan: Ipanema National Forest, Iperó.** Ministry of the Environment, Brasilia.

IBGE. 1992. **Technical manual of Brazilian vegetation** (Technical Manuals in Geosciences Series - Number 1). Rio de Janeiro: Fundaçao Instituto Brasileiro de Geografia e Estatistica - IBGE, 92 p.

IBGE. 2012. **Technical Manual of Brazilian Vegetation** (Technical Manuals in Geosciences Series - Number 1). Rio de Janeiro: Fundaçao Instituto Brasileiro de Geografia e Estatistica - IBGE, 2ed. 271 p.

INGRAM S.W. and NADKARNI N.M. 1993. Composition and distribution of epiphytic organic matter in a Neotropical Cloud Forest, Costa Rica. **Biotropica** 25(4):370-383

INGRAM S.W., FERRELL-INGRAM K. and NADKARNI N.M. 1996. Floristic composition of vascular epiphytes in a neotropical cloud forest, Monteverde, Costa Rica. **Selbyana** 17:88-103

ISHARA K.L., DÉSTRO G.F.G., MAIMONI-RODELLA R.C.S. and YANAGIZAWA Y.A.N.P. 2008. Floristic composition of a remnant of cerrado *sensu stricto* in Botucatu, SP. **Revista Brasileira de Botânica** 31(4):575-586.

JOANITTI S.A., CAVASSAN O. and WEISER V.L. 2010. Vascular epiphytism in a stretch of cerrado in the Legal Reserve of UNESP's Bauru Campus. *In:* **XXII Congress of Scientific Initiation of UNESP, 2010, Marilia, Brazil.** Proceedings of the XXII UNESP Scientific Initiation Congress. Marilia, Brazil: UNESP, 2010. p.18921895.

JOHANSSON D. 1974. Ecology of vascular epiphytes in West African rain forest. **Acta Phytogeographica Suecica** 59:1-136.

KERSTEN R.A. 2006. **Vascular epiphytism in the Upper Iguaçu basin, Paranà.** Doctoral thesis, Federal University of Paranà, Curitiba.

KERSTEN R.A. 2010. Vascular epiphytes - History, taxonomic participation and relevant aspects, with emphasis on the Atlantic Forest. **Hoehnea** 37(1):9-38

KERSTEN R.A. and KUNIYOSHI Y.S. 2009. Forest conservation in the Alto Iguaçu Basin, Paranà - Assessment of the vascular epiphyte community at different seral stages. **Floresta** 39(1):51-66

KERSTEN R.A. and SILVA S.M. 2001. Floristic composition and spatial distribution of vascular epiphytes in the coastal plain forest of Ilha do Mel, Paranà, Brazil. **Revista Brasileira de Botânica** 24:213-226.

KERSTEN R.A. and SILVA S.M. 2002. Floristics and structure of the vascular epiphyllous component in the Alluvial Mixed Ombrophilous Forest of the Barigüi River, Paranà, Brazil. **Revista Brasileira de Botânica** 5:259-267.

KIRA T. and YODA K. 1989. Vertical stratification in microclimate. In: H. Lieth & M.J.A. Werger (eds.). **Ecosystems of the world.** Vol.14b. Tropical Rain Forest Ecosystems. Elsevier. Amsterdam. p.7-53.

KOVACH W.L. 1993. **MVSP (Multivariate Statistical Package).** Kovach PLC.

KRESS J.W. 1986. A symposium: The biology of tropical epiphytes. **Selbyana** 9:1-22.

KRESS W.J. 1989. The systematic distribution of vascular epiphytes. In: U. Lüttge (ed.). **Vascular plants as epiphytes.** Berlin, Springer. p.234-262.

KROMER T., KESSLER M. and GRADSTEIN R.S. 2007. Vertical stratification of vascular epiphytes in submontane and montane forest of the Bolivian Andes: the importance of the understory. **Plant Ecology** 189:261-278.

KROMER T., KESSLER M., GRADSTEIN S.R. and ACEBEY A. 2005. Diversity patterns of vascular epiphytes along an elevational gradient in the Andes. **Journal of Biogeography** 32:1799-1809.

KRONKA F.J.N., NALON M.A., MATSUKUMA C.K., PAVÂO M., GUILLAUMON J.R., CAVALLI A.C., GIANNOTTI E., IWANE M.S.S., LIMA L.M.P.R., MONTES J., DEL CALI I.H. and HAACK P.G. 1998. **Areas of the cerrado domain in the state of Sao Paulo**. Sao Paulo, State Secretariat for the Environment, Forestry Institute.

LAUBE S. and ZOTZ G. 2003. Which abiotic factors limit vegetative growth in a vascular epiphyte? **Functional Ecology** 17:598-604.

LAUER W. 1989. Climate and Weather. In. H. Lieth & M.J.A. Werger (eds.). **Ecosystems of the world**. Vol.14b. Tropical Rain Forest Ecosystems. Elsevier.Amsterdam. p.54-75.

LEITE P.F and KLEIN R.M. Vegetaçao. In: **Geografia do Brasil, Regiao Sul.** v. 2. Rio de Janeiro: IBGE, 1990. p.113-150.

LEITE P.F. 2002. Contribution to the phytoecological knowledge of southern Brazil. **Ciência & Ambiente Magazine** 24:51-73.

LESICA P. and ANTIBUS R. 1990. The occurrence of mycorrhizae in vascular epiphytes of two Costa Rica Rain Forest. **Biotropica**. 22:250-258

LIMA R.A.F., DITTRICH V.A.O., SOUZA V.C., SALINO A., BREIER T.B. and AGUIAR O.T. 2011. Vascular flora of Carlos Botelho State Park, Sao Paulo, Brazil. **Biota Neotropica** 11(4): 173-214

LINSINGEN L., SONEHARA J.S., UHLMANN A. and CERVI A. 2006. Floristic composition of the Jaguariaiva Cerrado State Park, Paranâ, Brazil. **Acta Biologica Paranaensia** 35(3-4):197-232

List of Brazilian Flora Species. 2013. *In:* http://floradobrasil.jbrj.gov.br/ (Last accessed on 20/08/2013)

LUGO A.E. and SCATENA F.N. 1992. Epiphytes and climate change research in the Caribbean: a proposal. **Selbyana** 13:123-130.

LÜTTGE U. 1989. Vascular epiphytes: Setting the scene. *In.***Vascular plants as epiphytes** (U. Lüttge, ed.). Ecological Studies 79. Springer-Verlag, Berlin, p.1-14.

LÜTTGE U. 2004. Ecophysiology of Crassulacean Acid Metabolism (CAM). **Annals of Botany** 93:629-652.

MAACK R. 1968. **Physical geography of the state of Paranâ.** BADEP/UFPR/IBPT.

MACHADO R.B., RAMOS NETO M.B., PEREIRA P.G.P., CALDAS E.F., GONÇALVES D.A., SANTOS N.S., TABOR K. and STEININGER M. 2004. Estimates of area loss in the Brazilian Cerrado. **Conservation International** Brasilia, DF, Brazil.

MADISON M. 1977. Vascular epiphytes: their systematic occurrence and salient features. **Selbyana** 2:1-13.

MAGURRAN A.E. 1988. **Ecological diversity and its measurement.** Princeton University, Princeton,179 p.

MAMEDE M.C.H., CORDEIRO I. and ROSSI L. 2001. Vascular flora of Serra da Juréia, Iguape Municipality, Sao Paulo, Brazil. **Bulletin of the Botanical Institute** 15:63-124.

MANETTI L.M., DELAPORTE R.H. and LAVERDE Jr. A. 2009. Secondary metabolites of the bromeliaceae family. **Quimica Nova** 32(7):1885-1897.

MARTINELLI G., VIEIRA C.M., GONZALEZ M., LEITMAN P., PIRATININGA A., COSTA A.F., FORZZA R.C. 2008. Bromeliaceae of the Brazilian Atlantic Forest: species list, distribution and conservation. **Rodriguésia** 59(1):209-258

MAY R.M. 1975. Patterns of species abundance and diversity. *In.* **Ecology and evolution of communities** (M.L. Cody & J.M. Diamond, eds.). Belknap Press of the Havard University Press, Cambridge, p.81-120.

MEIRA M.S. 1997. **Spatial distribution of terrestrial Bromeliad populations in a forest and field mosaic.** Master's dissertation in Botany. Federal University of Rio Grande do Sul. Porto Alegre.

MENINI-NETO L., FORZZA R.C., ZAPPI D. 2009. Angiosperm epiphytes as conservation indicators in forest fragments: A case study from southeastern Minas Gerais, Brazil. **Biodiversity and Conservation** 18:3785-3807.

MORAN R.C. 1995. Clave para las familias de Pteridofitas. In: G. Davidse, M.S. Souza, &S. Knapp, (eds.). **Flora Mesoamericana. V.1. Psilotaceae to Salviniaceae.** D.F: Universidad Nacional Autónoma de México. Mexico. p.1-2.

MYERS N., MITTERMEIER R.A., MITTERMEIER C.G., FONSECA G.A.B., KENT J. 2000. Biodiversity hotspots for conservation priorities. **Nature** 403:853-858

NADKARNI N.M. 1981. Canopy Root: covergent evolution in rainforest nutrient cycling. **Science** 214:1023-1024.

NADKARNI N.M. 1984. Epiphyte biomass and nutrient capital of a neotropical elfin forest. **Biotropica** 16(4):249-256.

NAEEM S. 2003. Models of ecosystem reliability and their implication for the question of expendability. In.**The importance of species: perspectives on expendability and triage** (P. Kareiva & S.A. Levin, eds.). Princeton University Press, Princeton, p.109139.

NIEDER J., ENGWALD S. and BARTHLOTT W. 1999. Patterns of neotropical diversity. **Selbyana** 20(1):66-75.

NIEDER J., ENGWALD S., KLAWUN M. and BARTHLOTT W. 2000. Spatial distribution of vascular epiphytes (including Hemiepiphytes) in a Lowland Amazonian Rain forest (Surumoni Crane Plot) of Southern Venezuela. **Biotropica** 32(3):385-396

NIEDER J., PROSPERi J. and MICHALOUD G. 2001. Epiphytes and their contribution to canopy diversity. **Plant Ecology** 153:51-63.

NIMIS P.L., CASTELLO, M. and PEROTTI. M. 1990. Lichens as biomonitors of sulphur dioxide pollution in La Spezia (northern Italy). **Linchenologist** 22(3):333-344.

ODUM E.P. 1988. **Ecology**. Rio de Janeiro. Guanabara. 434p.

OLMSTED I. and JUAREZ M.G. 1996. Distribution and conservation of epiphytes on the Yucatan Peninsula. **Selbyana** 17:58-70.

PARKER G.G. 1995. **Structure and microclimate of forest canopies**. In. M.D.

Lowman & N.M. Nadkarni (eds.). Forest Canopies. Academic Press, San Diego, p. 73106.

PETEAN M. P. 2009. **Vascular epiphytes in an area of dense ombrophilous forest in Antonina, PR**. PhD thesis in Forest Engineering. Federal University of Paranà, Curitiba.

PETEAN M.P. 2003. **Floristics and structure of vascular epiphytes in an area of Altomontane Ombrophilous Dense Forest in Pico do Marumbi State Park, Morretes, Paranà, Brazil.** Master's dissertation, Federal University of Paranà, Curitiba.

PINTO A.C., DEMATTÊ M.E.S.P. and PAVANI M.C.M.D. 1995. Floristic composition of epiphytes (Magnoliophyta) in a forest fragment in the municipality of Jaboticabal, SP, Brazil. **Cientifica** 22:283-289.

PIRES J.S.R. 1995.**Anàlise Ambiental voltada ao Planejamento e Gerenciamento do Ambiente Rural: Abordagem Metodològica Aplicada ao Municipio de Luiz Antonio - SP**. Doctoral thesis. Federal University of Sao Carlos. Sao Carlos. Sao Paulo.

PIRES J.S.R. 2001.Diretrizes Conceituais e Metodológicas sobre a incorporação do Tema

215

Biodiversidade para o ZEE Brasil.In:**MMA/SDS. Methodological Guidelines for the Ecological-Economic Zoning of Brazil.** Secretariat of Policies for Sustainable Development, Ministry of the Environment, Brasilia.

PIRES J.S.R., SANTOS J.E. and PIRES A.M.Z.C.R. 2005. Bioregional Management. A conceptual approach to landscape management. In. Santos, J.E.; Cavalheiro, F.; Pires, J.S.R.; Oliveira, C. H.& Pires, A.M.Z.C.R..**Faces of Landscape Polysemy: Ecology, Planning and Perception**. Editora RiMa - FAPESP, V.I. Sao Carlos, Sao Paulo. p. 23-34.

ZERO REPORT. 2005.Update of the water resources situation report 1995 for the Sorocaba and Mèdio Tietê basin as a subsidy to the preparation of the basin plan. **Technical report n⁰ 80 401-205.** Sorocaba. Sao Paulo.

RIBEIRO J.F. and WALTER B.M.T. 1998. Phytophysiognomies of the Cerrado biome. In: Cerrado: environment and flora (S.M. Sano and S.P. Almeida, eds.). **Embrapa**, Brasilia, p.89166.

RICHARDS P.W. 1952. **Tropical rain forest - an ecological study**. Cambridge Univ. Press, Cambridge. 450p

RICHARDS P.W. 1996. **The tropical rain forest: an ecological study**. Cambridge: Cambridge University Press. 2ed. 600p

RICHARDSON K.A. and CURRAH R.S. 1995. The fungal community associated with the roots of some rain Forest epyphyte on Costa Rica. **Selbyana** 16:49-73.

RODERJAN C.V., GALVAO F., KUNIYOSHI Y.S. and HATSCHBACH G.G. 2002. The phytogeographic units of the state of Paranâ, Brazil. **Ciência & Ambiente Magazine**, 24:75-92.

ROGALSKI J.M. and ZANIN E.M. 2003. Floristic composition of vascular epiphytes in the strait of Augusto César, Seasonal Deciduous Forest of Alto Uruguai, RS, Brazil. **Revista Brasileira de Botânica** 26:551-556.

ROUSSE A. 1994. Xeric Bromeliads, **Journal of the Bromeliad Society.** 44(2):110- 117.

SAO PAULO (State). 2006. **Decree No. 51.381, of December 19, 2006.** Creates, in the Municipality of Anhembi, the Barreiro Rico Ecological Station, and makes related provisions. Poder Executivo, Sao Paulo, v. 116, n. 240, 20 dec. 2006. Section I, p. 1.

SAO PAULO (State). 2010. **Juruparà State Park Management Plan:** executive summary. Sao Paulo: Secretariat for the Environment: Foundation for the Conservation and Production of Forests in the State of Sao Paulo, 89p.

SCHEINVAR E. 1985. **Cactaceae. Flora Ilustrada Catarinense (CACT).** Herbârio Barbosa Rodrigues, Itajai.

SCHIMPER A.F.W. 1888. **Die epiphytische Vegetation Amerikas**. Jena, Verlag von Gustav Fischer.

SCHIMPER A.F.W. 1903. **Plant-geography upon a physiological basis.**Clarendon Press, Oxford, UK.

SCHÜTZ-GATTI A.L. 2000. **The vascular epiphytic component in the Salto Morato Nature Reserve, Guaraqueçaba - PR.** Master's dissertation, Federal University of Paranâ, Curitiba.

SETZER J. 1966. **Climatic and ecological atlas of the state of Sao Paulo**. Sao Paulo. International Commission of the Paranà-Uruguay River Basin.

SILLETT S.C. 1999. Tree crown struture and vascular epiphyte distribution in Sequoia sempervirens rain forest canopies. **Selbyana** 20(1):76-97.

SILVA J.M.C. and BATES J.M. 2002. Biogeographic patterns and conservation in the South American cerrado: a tropical savanna hotspot. **BioScience** 52:225-233.

SMA 2006. **Environmental Quality Report for the State of Sao Paulo.** Secretariat for the Environment. Sao Paulo.

SMITH L.B. 1962. Origins of the flora of Southern Brazil. **Contributions from the United States National Herbarium** 35(3):215-249.

SOARES-SILVA L.H. and BARROSO, G.M. 1992. Phytosociology of the arboreal stratum of the forest in the northern portion of the Mata dos Godoy State Park, Londrina-PR, Brazil. In. **Anais** do VIII Congresso da sociedade de botânica de Sao Paulo p.101-112.

SOARES-SILVA L.H., BIANCHINI E., FONSECA E.P., DIAS M.C. MEDRI M.E. and ZANGARO-FILHO W. 1992. Floristic composition of the arboreal component of the riparian forests of the Tibagi River, Paranà: 1. Doralice Farm - Ibipora, PR. *In:* **Anais.** Congresso Nacional sobre Essências Nativas, Sao Paulo. p.199-220.

SOTA E.R. de la. 1971. Epiphytism and pteridophytes in Costa Rica (Central America). **Nova Hedwigia** 21:401-465.

STANCATO G.C., MAZZAFERA P. and BUCKERIDGE M.S. 2002. Effects of light stress on the growth of the epiphytic orchid Cattleya forbesii Lindl. X Laelia tenebrosa Rolfe **Revista Brasileira de Botânica** 25(2):229-235.

STEEGE H. and CORNELISSEN J.H.C. 1989. Distribution and ecology of vascular epiphytes in lowland rain forest of Guyana. **Biotropica** 21:331-339.

TILMAN D. 1988. **Plant strategies and dynamics and structure of plant communites.** PrincetonUniversity Press, Princeton.

TILMAN D. and LEHMAN C. 2001. Biodiversity, composition and ecosystem processes: theory and concepts. In. **The functional consequences of biodiverdity: empirical progress and theoretical extensions** (A.P. Kinzig, S.W. Pacala & D. Tilman, eds.). Princeton University Press, Princeton, p.9-41.

TOPPA R.H. 2004. Structure and floristic diversity of different Cerrado physiognomies and their correlation with the soil at the Jatai Ecological Station, Luiz Antônio, SP. **PhD Thesis**, Federal University of Sao Carlos, Sao Carlos, 127p.

TRIANA-MORENO L.A., GARZÓN-VENEGAS, N.J., SANCHEZ-ZAMBRANO, J. and VARGAS, O. 2003. Vascular epiphytes as indicators of regeneration in intervened forests of the Colombian Amazon. **Acta Biològica Colombiana** 8:31-42.

TROPICOS. 2013. http://www.tropicos.org. (last accessed on 11/08/2013).

UHLMANN A., CURCIO G.R., FRANKLIN G. and SILVA S.M. 1997. Relationships between the distribution of phytophysiognomic categories and geomorphic and pedological patterns in an area of savannah (cerrado) in the state of Paranà, Brazil. **Archives of Biology and Technology** 40(2):473-484.

VELOSO H.P., OLIVEIRA FILHO L.C., VAZ A.M.S.F., LIMA M.P.M., MARQUETE, R. e BRAZÂO, J.E.M. (orgs.) 1992. **Technical manual of Brazilian vegetation.** IBGE, Rio de Janeiro, v.1.

VELOSO H.P., RANGEL FILHO A.L. and LIMA J.C. 1991. **Classification of Brazilian vegetation adapted to a universal system.** Rio de Janeiro: Brazilian Institute of Geography and Statistics (IBGE), Department of Natural Resources and Environmental Studies.

VERSIEUX L.M. and WENDT T. 2007. Bromeliaceae diversity and conservation in Minas Gerais, Brazil. **Biodiversity and Conservation** 16:2989-3009.

VIANA V.M. 1990. Biology and management of forest fragments. In. Brazilian Forestry Congress, Campos do Jordao. **Proceedings.** Brazilian Society of Forestry/Society of Forest Engineers, Curitiba. p.113-118.

VIANA V.M. 1995. Biodiversity conservation of tropical forest fragments in intensively cultivated landscapes. In: Interdisciplinary approaches to biodiversity conservation and land use dynamics in the new world.Belo Horizonte/Gainesville: **Proceedings.** Conservation International do Brasil/Universidade Federal de Minas Gerais/University of Florida, 1995. p. 135-154.

VIANA V.M. and PINHEIRO L.A.F.V. 1998. Biodiversity conservation in forest fragments. **IPEF Technical Series** 12(32):25-42.

VIANA V.M., TABANEZ A.A.J. and MARTINS J.L.A. 1992. Restoration and management of forest fragments. In. **Anais.** Congresso Nacional Sobre Essências Nativas, 2, Sao Paulo: Instituto Federal de Sao Paulo. p.400-407.

WAECHTER J.L. 1992. **Vascular epiphytism in the coastal plain of Rio Grande do Sul.** PhD thesis, Federal University of Sao Carlos, Sao Carlos.

WAECHTER J.L. 1998. Vascular epiphytism in a restinga forest of subtropical Brazil. **Revista Ciência e Natura** 20:43-66.

WALLACE B.J. 1989. Vascular epiphytism in Austro-asia. p. 261-282. In: H. Lieth & M.J.A. Werger, eds. **Ecosystems of the World.** V.14b. Tropical Rain Forest Ecosystems. Elsevier. Amsterdam.

WALTER H. 1986. **Vegetation and climate zones**. EPU, Sao Paulo.

WANNAZ E.D. and PIGNATA M.L. 2006. Calibration of four species of Tillandsia as air pollution biomonitors. **Journal of Atmospheric Chemistry** 53: 185-209.

WEAVER P.L. 1972. Cloud moisture interception in the Luquilo moutains of Puerto Rico. **Carribbean Journal of Science** 12:129-144.

WETTSTEIN R.R.V. **Aspectos da vegetaçao do Sul do Brasil**. Sao Paulo: Edgard Blücher, 1970.

WOLF J.H.D. 2005. The response of epiphytes to anthropogenic disturbance of pine-oak forests in the highlands of Chiapas, Mexico. **Forest Ecology and Management** 212:376-393.

World Checklist of Selected Plant Families 2013. http://apps.kew.org/wcsp/home.do (Last accessed on 15/08/2013).

ZOTZ G. and HIETZ P. 2001. The physiological ecology of vascular epiphytes: current knowledge, open questions. **Journal of Experimental Botany** 52:2067-2078.

ZOTZ G., HIETZ P. and SCHMIDT G. 2001. Small plants, large plants: the importance of plant size for the physiological ecology of vascular epiphytes. **Journal of Experimental Botany** 52: 2051-2056.

Printed by Books on Demand GmbH, Norderstedt / Germany